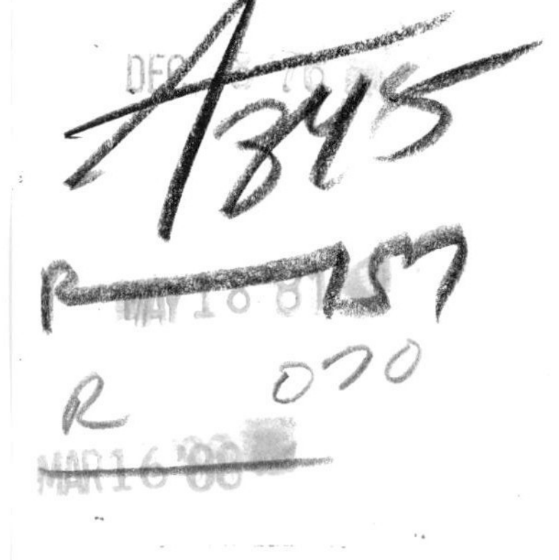

U.S. NEOCOLONIALISM IN AFRICA

50
YEARS
1924-1974

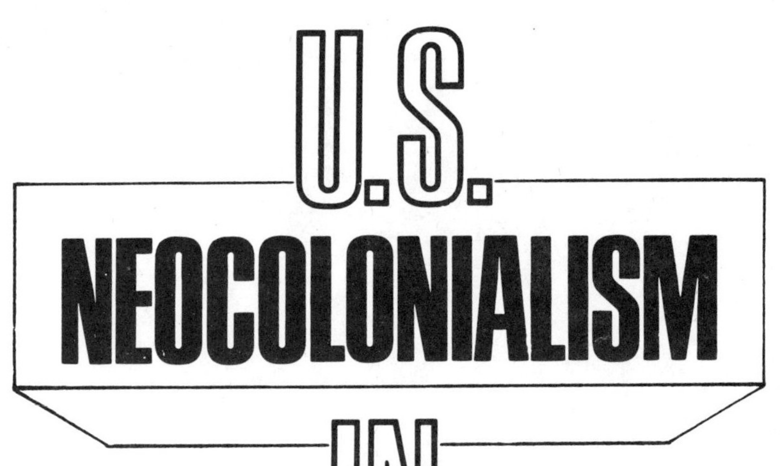

IN AFRICA

by Stewart Smith

INTERNATIONAL PUBLISHERS New York

First U.S. Edition by International Publishers, 1974
Printed in the United States of America

209

Library of Congress Cataloging in Publication Data
Seborer, Stuart J.
US neocolonialism in Africa
Includes bibliographical references
1. Africa—Relations (general) with the United States.
2. United States—Relations (general) with Africa. I. Title.
DT38.S4 301.29'73'06 73-87863
ISBN 0-7178-0392-9
ISBN 0-7178-0393-7 (pbk)

CONTENTS

FOREWORD

Africa is one of the main foci of struggle between the forces of national liberation and imperialism, one that is growing in importance. It has a special significance for the people of the United States because US imperialism, sustaining setbacks in other areas, more and more turns to Africa as a place in which to seek vast profits from the oppression and suppression of the peoples, from the plundering of immensely rich natural resources. It has a special significance because of the natural sympathy of the Black people within the United States for the African peoples' struggles and because of the Pan Africanists' efforts to divert that sympathy into providing unwitting assistance to the bitterest enemies and misleaders of the African peoples.

Knowledge of the real situation enhanced by the searchlight of Marxist-Leninist theory is necessary if the people of this country are to find the correct road of struggle in support of the African liberation forces and simultaneously in support of our own liberation from the exploitation and racist oppression of monopoly capital. Heretofore our sources of up-to-date knowledge about Africa have been severely limited. There have been exposés of particular situations such as the role of US corporations in South Africa or of US and other neo-colonialists in the murder of Lumumba. There have been broader historical sociological works and works of fiction providing deep insights.

We now have the kind of volume needed today: an all-round scientific study of US imperialism and Africa. The progressive American economist, Stewart Smith, has examined this in all four major dimensions: economic, political, cultural and military. We get from him in one place the total continent-wide picture of US corporate profiteering and expansion in Africa with necessary details for each major country and area. We get a deeper understanding of the flexible political strategy of US imperialism, its reliance everywhere on that reactionary exploiting class most capable of suppressing the masses, most amenable to making deals which subordinate national sovereignty to the interests of imperialism in general and US imperialism in particular, whether it be the fascist racists of South Africa, the Portuguese colonialists or the bureaucratic capitalist class in some newly independent countries. And we are provided with a clear exposé of the use of US advocates of Black capitalism and Pan Africanists within the general scheme of expansionism and domination.

Smith also provides comprehensive data on the continent-wide activities of the Pentagon, from massive support of the colonialists in Southern Africa to armament and technical training provided to selected reactionary governments and independent states, to collaboration with Israel as well as against the Arab countries of Northern Africa. He brings out the dual strategic of the Pentagon: suppression of the liberation movements within the continent and the use of African bases and resources as a reserve of growing importance in the global confrontation with Communism, that great liberating force which gains ever higher stature in the minds, hearts and practical organization of the African peoples as of all peoples of the world.

Smith writes from the vantage point of the Institute of World Economy and International Relations in Moscow where he is currently conducting research work. Relying mainly on US and other Western sources he also makes use of works of Soviet researchers not so readily available here. He wisely avoids the error of those armchair intellectuals who try to tell the progressive forces how to conduct their struggles.

I am certain that this volume has a big future as a weapon in the struggle of some 400 million African people for liberation and it is a *must* for all those people of the United States—white and black—who are concerned with this struggle and determined to end the crimes of US imperialism against the peoples of Africa. This book is a mighty contribution to all anti-imperialists in the United States and throughout the world in the struggle to develop worldwide support to the African liberation movements fighting for political and economic independence and complete victory over imperialism.

Henry Winston
National Chairman,
Communist Party, USA

PREFACE

This brief survey aims at a synthesis of US relations with Africa by concentrating on the continent as an organic part of world development. It covers the major spheres as they relate to each other in the period from the mid-fifties to the present day (cut-off date 1971, in a few cases 1972). The study examines the primary political aims and policies of US ruling circles and the slower moving economic monopoly interests, as well as some important military aspects. It discusses the social strata affected, the ideologies based on these, and the world influences involved. In sum, a qualitative measure is sought of the totality of strength and pressure brought to bear by the world's biggest imperialist power on the African continent.

From the standpoint of the United States, the evolution of its postwar policy toward Africa is outlined with particular relevance to its global foreign policy, internal processes within the bourgeoisie, and the special role of the American people, particularly the Blacks—their contribution to African liberation and its retroactive effect on the Black liberation struggle in the United States.

How and to what extent are Washington's efforts to attain prime political aims by maximizing imperialist strength through NATO and other allies, flexibility in operating in different spheres, and through joint institutions and actions, offset by the contradictions and weaknesses of the imperialist powers in their relations to one another and, even more important, vis-à-vis the African peoples and their Socialist and other world allies? To what extent is US policy influenced by decision-makers' conceptions of strength, their estimates of the ebb and flow of the African struggle for political and economic independence, and the bias of the cold war to intervention? These are questions posed in the political sphere.

A radical treatment of the economic basis (like the other spheres) as a tightly knit complex of related sub-categories—in contrast, for instance, to a liberal "factor theory" with its eclectic approach—disposes one to seek the dialectic process of interaction, in this case, between the export of capital (and inflow of profits), aid and trade. And one might justifiably anticipate such an inter-relationship, in turn, to be closely correlated with political aims. The resultant, in fact, reveals a definite—even if not abstractly neat—"nesting" pattern. At this point, furthermore, US political-economic foreign policy, by its nature, flows over into the African social sphere.

Although Washington may "tilt" toward the dominant classes of outlived formations and those with former colonial ties to imperialism (feudal and comprador in the North, tribal and feudal in Subsahara, plantation owners and capitalists in the South), does it also show a keen interest in promoting the developing bourgeoisie? And is it concerned as well in influencing the young African working class? These are disturbing questions of US policy for both Africans and Americans alike.

Since the capitalist socio-economic stage is not evolving classically nor inevitably in Africa, the force of ideas and influence of world forces and events in the ideological struggle carry particular weight. Quite aware of this, Washington devotes considerable efforts and funds to divert Arab and black Africa from the non-capitalist path. The study describes the substance, forms and methods, and influence of subversive US imperialist propaganda—the spreading of anti-Communism, promotion of nationalism and tribalism, and efforts to rationalize US ties with the southern racists and colonialists. Yet, one could hardly expect "hard-headed" American politicians and businessmen to permit the war of ideas to replace the big stick in US policy.

Since political-military and strategic considerations do loom so large in Washington, certain aspects of them must be examined more closely, for example, in tropical Africa—the disproportionate US military aid to Ethiopia, and the US role in the conflicts in the Congo, now Zaïre; the colonial and racist wars and suppression in southern Africa; and the growing US active involvement since Suez in the Afro-Arab states, which are inseparably part of the regional Middle East conflict. Here, one cannot opt out of dealing with the complicated problem of how US official strategy and tactics affect general tendencies, and the prospects for either armed hostilities or political settlement.

The concluding chapter essays a brief overview of major currents among bourgeois ideologists and policymakers which affect US policy toward Africa and their underlying conceptions of strength, the Nixon doctrine and its implications, and perspectives for the 1970's. This applies above all to the Israeli-Arab complex in the North and the southern Africa settler complex—the two strategic centers of US imperialist policy on the continent.

Briefly concerning sources, we have sought to use official documents and reputable bourgeois publications and studies as much as possible. We also have turned to the still small, but growing number of Soviet and other Marxist-Leninist studies, which are providing, in the final analysis, a more profound understanding of the processes at work. In this undertaking, we have had the warm cooperation of the librarians of various Moscow libraries, particularly at the Institute of World Economy and International Relations.

In writing this book, the author is deeply indebted to many people. He wishes to express particular appreciation for the inspiration of Gus Hall and Henry Winston; the personal interest of James S. Allen and I. G. Needleman; the helpfulness of Academician N. Inozemtsev, Director of the Institute of World Economy and International Relations, Deputy Director E. Primakov, and Scientific Secretary Z. Litvin; and the lively discussions with his colleagues, notably O. Bykov, Ya. Etinger, M. Gelfand, V. Kaplan, D. Maclean, Yu. Melnikov, G. Mirsky, V. Rymalov, Y. Tarabrin and M. Voztchikov, of the Academy of Sciences, U.S.S.R.

In closing, I wish to thank more than words can do my brother Oscar, without whose unfailing support and encouragement this book could not have been written.

Stewart Smith

Moscow, July 1972

To Amilcar Cabral and all dedicated fighters for national liberation and Socialism

I. INTRODUCTION

1. AFRICA AND WORLD MOVEMENT

Today, one needs no elaborate justification for treating continental Africa—and consequently US relations to Africa—as a whole. But a few short years ago, the continent was customarily dealt with in dismembered form. North Africa, and particularly such countries as Egypt,[1] Libya, the Sudan and Somalia, were frequently excluded—on the grounds that they are part of the Middle East. South of Sahara, or tropical Africa, frequently has been dealt with separately. And South Africa also is excluded frequently on the grounds that it has a relatively developed economy[2] as compared with the rest of the continent. However understandable many of these exclusions may be from certain historical, and for some present-day statistical, reasons, much of the close interconnection and dynamics of Africa today would be lost by leaving out either northern or southern Africa. The role of the Arab Republic of Egypt (ARE)[3] in Africa is an especially vital one, quite apart from its identification

[1] Egypt especially has been historically treated separately from the rest of Africa and often regarded anthropologically as Asiatic, or even as "white" or European (Arnold Toynbee's *Study of History*). In part, this may be due to the great influence which the Nile culture had on the later European civilization. On the other hand, part of the rationalization for the slave trade and Negro slavery, particularly in the 19th century, called for differentiating out and relegating the Negroid peoples of tropical Africa to a backward and even sub-human level. So it may not be mere coincidence, for example, that the separate science of Egyptology flourished during this period. (See especially on this point *The World and Africa* by W.E.B. DuBois, N.Y., 1965).

[2] This, for example, is UN usage.

[3] The official name as of September 2, 1971, replacing the designation United Arab Republic dating from the Union of Egypt and Syria in February 1958 and continued to be used in Egypt after the dissolution of that union in September 1961.

with the Middle East and the Arab world as such. Furthermore, the fact that South Africa is the keystone in colonial and racist Africa, intimately connected economically and politically with Rhodesia, Mozambique, Angola and Namibia (South-West Africa), cannot be omitted without artificially truncating one's understanding of the continent.

This vast continent of over 300 million people, with its own millennia of many-layered societies, has been most strongly influenced from abroad through military invasion, political domination and economic ties. Of the imperialist powers now active there, the United States came last upon the scene, and to some extent—at least by virtue of its global position—stands on top of the pyramid of foreign influence. To understand US policy, therefore, one must deal with it and relate it—as does Washington—particularly to such important recent and continuing influences as those of the European imperialist powers.

Almost four centuries ago Europe began developing her commerce and then industry, which was to give her economic, political and military supremacy in the world. The three economically underdeveloped continents of the present day helped provide sources of primitive accumulation. In the period of exploration and commerce, the search for precious metals, spices and luxuries to supply the needs and comforts of the burgeoning bourgeoisie of the developing European national states led to the gold and silver mines in America, the tropical products of the East Indies and the slaves of Africa.[1] The plunder from these continents provided both wealth and stimulus for further commercial expansion, which contributed to the development of capitalist manufacture and industry in Europe and America, the colonial system, and the consequent exploitation of labor both at home and abroad.

The further development of the industrial revolution in Europe and North America, giving rise to increased demand for raw materials, markets and sources of higher rates of profit, led eventually from the capitalism of "free competition" to monopoly, from colonialism to imperialism. This great expansion of the late 19th and 20th centuries, which brought the capitalist national economies into a world capi-

---

[1] See Karl Marx, *Capital*, Vol. I, Ch. XXXI "Genesis of the Industrial Capitalist".

talist system—analyzed by Lenin in the classic *Imperialism, the Highest Stage of Capitalism*—led to economic and political conflict, war on a world-wide scale, and general crisis.

The emergence of Soviet Russia as the first successful working-class state, its industrialization and development as an attractive international political and social force, the decisive victory over fascism won by Socialist and other democratic forces in World War II, the further breakaway from capitalism of Eastern European and Far Eastern countries in the postwar period, as well as the disintegration of the colonial system dealt major blows to world imperialism.

The successes of national liberation in Africa, which reached a crescendo in the second postwar decade, came, to be sure, as a direct result of the heroic efforts of the oppressed and exploited African peoples—as the fruition of decades of self-sacrificing and bloody struggle for the right to govern themselves and to enjoy the proceeds of their own land, resources and labor. Within their global context, the victories embodied long years of anti-imperialist struggles led by the international working class and its first Socialist state. Particularly decisive was the defeat of fascism and militarism in Europe and the Far East, which also witnessed the weakening in Asia and Africa of the classical colonialist powers—Britain, France, the Netherlands, Belgium and Portugal.

It is against the background of a changing balance of forces between the Socialist and capitalist world systems in the mid-1950's that the victories of the national-liberation movement, culminating in the 1960 "Year of Africa", could be won.

The decade which followed witnessed both ebb and flow in African struggles and progress. The United States, which began taking a major interest in African affairs during this period, contributed no small part to the ebb. Why, how and to what extent are questions which especially concern us.

2. TREATMENT

The forces acting today in Africa include, to be sure, the social and economic patterns of traditional pre-colonial and colonial times, e.g., feudalism and tribalism. For historical layers are not obliterated by a neat superimposition of the succeeding period but linger on, intermixing with the new and leaving their imprint. However, it is argued here, that their importance is secondary and subordinate, on the whole, to more recent African and particularly world socio-political forces which seek either to promote or to retard African political and economic independence.

The Marxist-Leninist materialist conception of history, with its systematic approach, has given us an indispensable tool for understanding and analysis of the social sciences[1] and the world as a whole. The complex of political, economic and other factors which are effective in a given society are frequently separated out, at least for purposes of analysis. In life, of course, they are clearly interwoven and interacting,[2] and when mutually supporting they provide stability and strength. Sometimes, however, when there exist serious lags—the heritage of previous or incompleted historical processes and periods, there arise contradictions generating instability and weakness. Although imperialism as such may not have given rise to all of the latter in Africa, it has sought to perpetuate or take advantage of many of them in its own interests and to the detriment of the African people.

Of the major forces affecting the United States and Africa in the 1960's, African political changes have been of primary importance. This is reflected in the continuing, sharp revolutionary struggle for political independence from imperialism and colonial rule, without which there would be no great impetus for economic and social change. Economic development, a much slower evolutionary process within the framework of sovereign power, has become a prime goal of the young states seeking to correct their distorted, primary-producing economies inherited from colonialism in order to ensure political and economic independence and to improve living standards. Social forces and

---

1 In one form or another, the concept of closely interrelated social science study is more and more being recognized by non-Marxist writers. Thus, for example, see reference to the concept of "interdisciplinary" study, or the "standpoint of social science as a whole" rather than any single social science in *Three Worlds of Development* by J. L. Horowitz, N.Y., 1966.

2 See Engels' letter to F. Mehring, 14 July 1893, in Karl Marx and Frederick Engels, *Selected Works*, Vol. II, Foreign Languages Publishing House, 1951, pp. 450-54.

changes are linked largely with the chosen path of political and economic development, and especially concern questions affecting the means of production, e.g., nationalization, agrarian reform, the public sector, foreign investment. New leaders, parties, classes and social relations which are emerging come into conflict with opposing and retarding forces—both domestic and foreign. In the ideological sphere, of key importance are the social ideas, concepts and theories motivating and resulting from the above conflicts, struggles and changes; the educational, religious and cultural influences—European and African, national and international, capitalist and Socialist. Finally, of critical importance in safeguarding or undermining the independence of the young states is their military strength. In addition to the level of organization, training and equipment, this includes, for example, questions of political and ideological allegiance, and social composition of the armed forces. In sum, and in the broadest sense, however, the strength of the African states consists of the totality of these factors taken in conjunction with that of their continental and world allies.

Recognition of the interdependence of these factors is of significance both to Africa as well as to US imperialism. For the former especially it has many implications. It raises such questions as to whether serious strides can be made within the socio-economic sphere while imperialism continues its political and economic domination? Or, the net advantage to be gained from minor economic advances if political independence is jeopardized, e.g., through close ties with imperialism? Or, whether short-term economic gain is in principle—and therefore in the long-term interest—justified if African or world social progress as a whole is held back, e.g., in southern Africa, the Middle East or Vietnam? Or, whether the "luxury" of interpreting "non-alignment" to mean succumbing to anti-Communism does not thereby deprive the new state of its natural allies and its potentially superior ideological strength? The obverse of these and related questions, it may be inferred, is taken into account by the makers of imperialist policy.

US imperialism, on the whole, apparently sensing that a systematic, international approach to the contemporary world scene works to its disadvantage, generally has sought to fragment and deal separately with these categories. To take one example, at the second UNCTAD conference held in Delhi in February-March 1968, the US representative, Eugene Rostow, insisted that trade be considered independent of political and social questions. On this basis, Washington sought to justify its open support for the participation in the conference of South Africa, the main pillar of apartheid and racist rule. Washington, also understandably, opposed any reference by the conference to its aggressive war in Vietnam.

Viewing the forces at work in Africa today, we are concerned with the recent past and present, but are also looking ahead to the implications for the short- and long-term future. Since movement and impulse to progress is the result of the resolution of the struggle of opposite and conflicting forces, its direction and speed correspond roughly to the changing correlation of those forces, e.g., with respect to the African struggle against imperialism for economic independence and greater political security in the independent states, and against colonialism and racist rule in the south. Since the overall strength of the antagonists includes both subjective and objective factors, much of the speed of movement depends on, their ability to convert subjective into objective force. Furthermore, in view of the great importance of international factors both for Africa and for imperialism, the effective strength which the imperialist powers—not least of all the United States—can bring to bear may be critical.

US imperialism recognizes that it cannot allocate a disproportionate share of its global strength to one region without running big risks elsewhere. It therefore seeks to break out and develop strategically the relationship of forces, including its own, into a pattern of strength which will work in its own favor. US imperialism, however, possesses varying elements and degrees of strength and weakness in different spheres, e.g., economic, political and military.

Recognition of this in US official and unofficial circles has led in recent years to greater attempts at an analysis of these relationships as well as at their measurement.[1] This has arisen not only from the desire to attain greater performance for each dollar expended but also from the awareness

---

[1] Thus, Herman Kahn and A. J. Wiener, of the Hudson Institute, find that power and influence are multi-dimensional concepts which include military capacity, size, wealth, geographical position and less precise notions such as stature, prestige, culture. For the quantitative

(probably in no small measure as a result of the rising costs of, and negative returns from, the Vietnam war) that US funds also have certain limits.

Thus an immediate aim of measurement is frequently to attain a better apportionment of US dollars in economic and military aid programs,[1] which are considered among the principal instruments of the US foreign policy in the Third World,[2] whereas "the political and psychological factors are least amenable"[3] to quantitative measurement. The emphasis of American imperialism on economic and military forces may not be accidental, in view of its weakness and vulnerability in the political, social and ideological spheres. Hence the great emphasis on the former two spheres in this study.

Imperialist policies and actions are here grouped into five spheres for analysis purposes. This, however, is not considered a final framework for neatly systematizing and measuring all forces, which may be tangible or not, and frequently are fragmented or even concealed. Since certain quantitative aspects of US relations with Africa, for example, in economic matters—which are basic in long-term historical development—are subject to measurement, they will be dealt with statistically. However, other factors from other spheres —which may be codeterminant, and perhaps even of cardinal importance particularly in the short-run period— may be qualitative. Our approach will seek to take into account both qualitative and quantitative factors in trying to arrive at a better understanding of some of the processes and trends in US-Africa relations in the 60's and early 70's.

---

measurement and comparison of the role of various countries to the end of the century, the authors employ 3 indices—population, GNP and per capita GNP (see *The Year 2,000*, N.Y., 1968).

[1] Thus, a recent RAND study seeks to determine the relationship among economic growth, redistribution of wealth and the role of force; and attempts to measure quantitatively US government military and economic progress and to determine how much money is required to keep a country within the US orbit as compared with the cost of its loss. (*US Policy and the Third World* by Chas. Wolf Jr., Boston, 1967).

[2] Ibid., p. 184.

[3] Ibid., p. 20. The critical instrument of US foreign policy in the underdeveloped world, according to another study, has been the military, then comes government aid followed by investment (*Imperial America, The International Politics of Primacy* by Geo. Liska, Baltimore, 1967, pp. 83-84).

II. CONTOURS OF THE US RELATIONSHIP

In close interconnection with Europe, the US relationship to Africa has involved several major socio-economic and political forces, including certain segments of the US bourgeoisie and the government apparatus, on the one hand, and the broad masses of Americans, particularly Blacks,[1] on the other. These have had not only contradictory interests and aims, but also different degrees of influence, with the monopolies on the whole predominating. But that is not to say that strength relationships have not altered in various periods and may not be expected to do so again in the future, especially in the context of world forces.

1. LEGACY AT THE END OF THE WAR

Briefly, although two centuries of the slave trade centered in Liverpool, and was a factor in England's capital accumulation which promoted the textile manufactures of Manchester, the end goal of the traffic, in the final analysis, was the New World. Here not only slave dealers and

[1] In place of "Negro" (termed a "slave word" by some), "Black" and "Afro-American" became for many the preferred designations. This was not merely semantic but clearly related to land, history, social and cultural ties. The New York Negro Teachers Association changed its name to the African-American Teachers Association. The late Ralph Bunche, Under Secretary for Special Political Affairs in the United Nations now used "Black" as often as "Negro". C. Eric Lincoln, the sociologist at Union Theological Seminary, uses "Black" when talking to young people, and "Negro" when addressing those past 40.

Southern planters reaped economic benefits, but also the slave runners and rum distillers of the pious bourgeoisie of New England. Small wonder that slavery was recognized and anchored in the US Constitution. At the expense of Africa,[1] enslaved Afro-Americans became the main source of labor power used in the South, which also provided raw materials and stimuli for both manufacturing and shipping in the North until the contradictions and struggle between the two regions for national supremacy resulted in the Civil War.

A legacy of slavery, moreover, was the economic and social discrimination against the Negro, both in legislation and practice, and the propagation of racist ideology. These have continued in the United States (and, of course, in colonial Africa) to the present day. They constituted no insignificant weapons of the rulers of America to divert, divide and weaken the working population, and to derive surplus profits from both white and black labor.

Another early, if minor, US direct relationship with Africa was the settling in the 1820's and 1830's of a small group of freed Negro slaves sent out by American "colonization societies" to establish themselves on the coast of Liberia and eventually colonize the interior. The handful of emigrants represented a nostalgic attempt to reply to domestic slavery and racism by turning back the wheel of history through a return to Africa. Even this small initial trickle, however, dwindled in the face of the prospect of freedom and equality in America through the growth of the Abolition movement before the Civil War.

The abolition of slavery and the emancipation of four million Negroes during the Civil War acted not only as a catalytic political force and manpower reserve in the struggle against the southern landowners, but also had its repercussions in Africa—the halting of the further bloodletting of its population. But with the coming of the epoch of imperialism the exploitation of that continent took other forms, in which the United States, however, played a lesser

---

[1] It has been estimated by W.E.B. DuBois (*The Negro,* N.Y., 1915, pp. 155-56) that the American slave trade meant the elimination from Africa of at least 60 million Negroes—about 50 million dying in the process, either in Africa or en route to the free world. See W. A. Hunton, *Decision in Africa,* N.Y., 1957, pp. 16-17.

role. Thus, it was not until the beginning of this century that small amounts of US capital, for example, found their way into the Congo, Liberia, Ethiopia, South Africa and Rhodesia.

In Liberia, over a century after its founding as a supposed haven for freed slaves, the descendants of these settlers—Americo-Liberians as they are called—constitute only about one percent of the country's inhabitants. However, they have become a ruling caste governing the overwhelming majority of the approximately two million Liberian population. These consist of African tribes occupying the vast interior, with their traditional tribal organization and village life. In the early part of the century, the economic prospects of the country, which lay in its climate, soil and labor, were eyed both by British and French colonialism, but were protected by the US government for American monopolies.

Thus, in 1926 the Firestone corporation undertook to develop extensive rubber plantations using the cheap labor ensured by a government which a League of Nations investigation in 1931 charged with practising widespead forced labor and semi-slavery. As a consequence, the leaders of the Liberian government were forced to resign, but international control of a country which Washington had come to regard as its "special responsibility" was warded off and the American rubber monopolies continued their operations with minor modifications. By the end of World War II, the United States was consolidating the country as a commercial and major strategic naval outpost in West Africa.

A new phase began during World War II, when the US government with its landing of troops in 1942 widened its influence in North Africa. And with a dollar "mortgage" on the British economy through lend-lease shipments, it expanded substantially its small volume of trade with tropical Africa. This took place mainly through US imports of raw materials and foodstuffs shipped on the United Kingdom account from British colonies—copper, chrome, asbestos, graphite, sisal, palm, peanuts and cocoa. Nevertheless, this volume of trade was still small, and the bulk of US commerce continued to be with South Africa. The enlarging US foothold on the continent as a whole in the course of the war resulted essentially from the weakening position of the European colonial powers, due partly to the direct war strain on them, as well as the participation of the

African people in the war in the hopes of achieving independence.

US governmental organization prior to the close of the war also reflected the changing relationship of the European colonial powers to Africa, as well as the broadening horizon of American monopoly interests. Thus, for many years before the Second World War, Department of State officials concerned with European affairs also had handled the occasional African matters that came up. In 1937, action responsibility for much of Africa was given to the Near Eastern Division, within which a separate Office for Africa was established—significantly in 1943. This paralleled the actual course of US military operations, as well as foretelling the increasing postwar interest of American monopolies in this oil-rich region.

The American Blacks' contribution to African independence since the turn of the century centered largely about Pan-Africanism, which sought to mobilize world opinion against colonialism in Africa and race discrimination in the United States. W. E. B. DuBois, the moving spirit, went so far at the Pan-African Conference in 1900 as to declare that "the problem of the 20th century is the problem of the color line".[1] The movement's social composition was relatively well-off educated professional and business men, who concentrated on writing and speaking on racial lines, rather than organizing on class lines. The First Pan-African Congress held in Paris in 1919 called upon the colonial powers to halt slave and forced labor, abolish corporal punishment and draw Africans into the government of the colonies. Although four Congresses were held prior to World War II,[2] it was the Fifth Pan-African Congress held in Manchester, England, in 1945 which, in the postwar world balance of forces, began to exert a marked influence. DuBois presided and participating were men who were soon to become outstanding African leaders, such as Kwame Nkrumah, Jomo Kenyatta, Peter Abrahams and George Padmore.[3]

---

[1] Quoted in *Africa, the Politics of Unity* by Immanuel Wallerstein, N.Y., 1967.

[2] The Second Congress (London, Brussels and Paris) in 1921, the Third Congress (London, Lisbon) in 1923, the Fourth (New York) in 1927.

[3] See "Africa and World Peace" by W.E.B. DuBois in *Political Affairs*, February 1968.

2. POSTWAR PERIOD THROUGH "YEAR OF AFRICA"

Africa's place in the postwar disintegrating colonial system can be historically divided into two distinct periods. *First*, until the late 1950's, when most of the African peoples were actively fighting to break centuries-old colonial ties in order to achieve political independence. During this period, US imperialism was usually in the background, but, on the whole, gave close support to its colonial allies in a determined but losing battle to prevent or delay the realization of the historical process of national liberation. *Secondly*, since the late 1950's, when most of Africa rapidly achieved sovereignty within a few short transitional years and then Africans began to play a growing role on their own continent and in international affairs within a changed world balance. This compelled the major metropolitan powers, with US encouragement, generally to adopt more indirect forms and methods of influence and domination. It also gave US imperialism more room for independent maneuver and a greater role in organizing reaction on a continent in flux.

IMMEDIATE POSTWAR YEARS

With the European metropolitan countries still holding the dominant political and economic positions in their colonies, it is not surprising that US economic (to all intents and purposes, big business) interests remained both absolutely and relatively small, especially in those countries, and were concentrated mainly in "independent" South Africa and Liberia. They represented only about 2% of US total private investment abroad and about 5% of its total world trade. However, they continued under the umbrella of, and therefore with a general stake in, the political "stability" provided by European colonial rule, apart from those indirect US monopoly interests and links with colonialism through ties with the metropolitan countries themselves.

Furthermore, from the earliest postwar years, the global ambitions of US imperialism to keep world capitalism intact under its increased domination paralleled the efforts of its European allies to prevent the collapse of their African colonial empires. The US contribution consisted, in part, in expanding its own independent military presence in

North Africa and the Mediterranean, as well as maintaining a joint force ensconced in NATO. This, too, was garbed in the cold war terms of "stopping Communism" and "defense" against an alleged threat from abroad. The real target, however, as is evident from official sources—and, more important, from events themselves—was the indigenous national-liberation movement, which represented a threat to imperialist interests.

Thus, joint imperialist politico-military objectives stressed the importance of military bases in keeping North Africa, "flanking both the NATO area and the oil fields and communications of the Near East",[1] oriented to the West. In the same vein, the "special political and military interests"[2] of the United States and its NATO-recognized sphere of responsibility in North Africa included military supply lines and bases in Libya, Morocco, Tunisia, Ethiopia and Liberia.

In the case of Subsaharan Africa, the direct imperialist interests were even less veiled. For, as a Senate Foreign Relations Committee study had to admit with respect to President Truman's Point 4 program (1949-52), Subsaharan Africa "had no immediate strategic significance".[3] Nonetheless, actual political and economic interests did exist and also required protection. Thus, NATO left military activities for Subsaharan Africa, "strategically" important to the United States as a source of "human and natural resources" and "vital supplies of essential materials", to the European NATO countries. Responsibility for southern Africa was left in the reliable hands of the racist Union of South Africa.

Although the United States historically inclined to indirect imperialist penetration and was at times embarrassed by the colonial policy of its allies, nevertheless, US ties tended to support the latter and thereby to preserve their African colonial empires. US efforts were not limited to, or essentially in, the military sphere. The European powers, already on the ground as colonial overlords, were directly engaged in suppressing the African anti-colonial move-

---

[1] *The Department of State Bulletin,* April 18, 1960, Assistant Secretary for African Affairs Satterthwaite, p. 607.

[2] Ibid.

[3] US Senate, Committee on Foreign Relations, *United States Foreign Policy in Africa,* October 23, 1959, Washington, p. 49.

ments—from Egypt to Kenya, from Algeria to Madagascar.[1] The US role in this connection consisted mainly in providing its NATO allies with financial and economic assistance, as well as military supplies and equipment. This—in net effect—division of labor among the imperialist powers was not lost upon the African peoples when, together with world progressive forces, they undermined and destroyed the colonial system.

PERIOD OF TRANSITION THROUGH "YEAR OF AFRICA"

Only after a changing world balance of strength, largely as a consequence of violent struggle and defeat at the hands of national-liberation forces in Asia and Africa, did the major colonial powers in Africa—first Britain, then France and Belgium—reluctantly abandon colonialism.

The changing correlation of forces to the disadvantage of imperialism also affected the position taken by US imperialism in such key struggles as were waged in Egypt, Algeria and later in the Congo.

Whereas *Britain*, for example, during the 1950's was still seeking to stifle the African anti-colonial movement in Kenya and Egypt by force of arms, US policy was concentrating mainly on applying various forms of non-military pressure. Thus, it was as a result of unacceptable Washington demands that the drawn-out US-Egyptian arms negotiations came to nought in 1955. Furthermore, the following year witnessed Dulles' gross miscalculation in withdrawing the Aswan Dam offer. However unsuccessful[2] these non-military pressures turned out to be, Washington did not feel it expedient to associate itself with such desperate military actions as the Anglo-Franco-Israeli Suez aggression in 1956. When this ended in a political fiasco, moreover, Britain also recognized that concessions to African sovereignty would have to be made. For with the emergence of a world Socialist system, the monopoly position of the capitalist world had been broken and arms, political support, mutually

[1] See *Les Damnés de la Terre* by Frantz Fanon, Paris, 1961. Translated as *The Wretched of the Earth,* New York, 1965.

[2] These US foreign policy measures, acknowledged Secretary of State Rusk, "all adversely affected the US position". *The Department of State Bulletin,* July 1, 1963, p. 24.

beneficial trade and economic assistance for development were now becoming available from the Socialist states.

In short order thereafter, independence was attained in tropical Africa—first in British West Africa by Ghana (1957), Nigeria (1960), and Sierra Leone (1961). Then in British East Africa by Tanganyika (1961), Uganda (1962), Kenya and Zanzibar (1963), Malawi and Zambia (1964). But this was done not without preparatory steps being taken by Britain, which made her the first major colonial power in Africa to embark broadly on a neocolonialist course. The new sovereign states were greeted in the United States by genuine enthusiasm on the part of the American people, by quick diplomatic recognition from Washington, and expanding governmental and monopoly ties.

In the midst of this transition, moreover, fearing that white supremacist intransigence in the remaining stronghold of colonial Africa, in which British imperialism had a formidable economic stake and to which it was linked by strong political ties, might well impel the entire national-liberation movement to more revolutionary action, Prime Minister Macmillan admonished the Union of South Africa parliament in February 1960: "The wind of change is blowing through the continent ... the growth of national consciousness in Africa is a political fact and we must accept it as such."[1] Indeed, the great issue in the present period, according to the Prime Minister, was which alternative path the peoples of the underdeveloped countries would take, and the black continent, he grandiloquently declared, might well hold "the precarious balance between East and West".

Recognizing that the balance of political power in the world was altering—this was especially evident in the United Nations—US foreign policymakers welcomed this call for a change in tactics on a continental scale. Masked colonialism could then be paraded to Africa and the world as imperialism having undergone a metamorphosis. Even though, as could have been foreseen, only its stripes had changed.

French colonialism also was overtaken by struggle and defeat in the 1950's. In withdrawing her troops from Southeast Asia to Africa, where most of her colonies were located, to crush Algerian resistance, France again became embroiled

---

[1] *The New York Times*, February 4, 1960.

in a large-scale bloody war. This once more began draining her military and economic strength, as well as politically weakening the French ruling class vis-à-vis the French people who strongly opposed this new "dirty war".

Without US support during this period, e.g., Washington's assistance to France in obtaining $500,000,000 in international loans and credits during 1958 alone, it would have been difficult for the latter to prosecute the war against the Algerian people for 7½ years. Recognition of this by the Algerian National Liberation Front[1] during its bitter struggle, which cost losses estimated up to one million lives, undoubtedly influenced post-liberation Algerian-US relations.

From a global viewpoint, US policymakers were concerned over the political and military consequences of the long-drawn out colonial war. Politically, it was exacerbating the relations of France—and, as a result, those of the supporting imperialist powers—with the developing countries. By exposing the antithesis between imperialism and national liberation, it was impelling the anti-colonial forces to seek greater assistance from their mutual allies, the Socialist states and the international working class. Consequently, President Eisenhower—perhaps not without Washington's ambitions in North Africa in mind—was urging upon de Gaulle the US-favored strategy of decolonization.[2]

As a result of pressures—mainly by North African liberation movements—upon France, Morocco and Tunisia achieved independence in 1956. Moreover, by 1958 French imperialism, seeing the handwriting on the wall, reluctantly switched over to the new colonialism in West and Equatorial Africa and the Malagasy Republic. The autonomy granted, however, did not include freedom to determine internal and external policies and relations. Key functions such as finance, defense and foreign affairs, rested in the

---

[1] See *Annuaire d'Afrique du Nord,* I, Paris, 1962.

[2] In a meeting with de Gaulle in September 1959, for example, President Eisenhower indicated that the United States could support France more fully in the United Nations, rather than abstaining as in 1958, if the French took "some prior constructive action respecting Algeria that would prove acceptable to world opinion—or at least present a full explanation of their point of view". Dwight D. Eisenhower, *The White House Years, Waging Peace 1956-61,* N.Y., 1965, pp. 429-30.

safe hands of the French Community, that is of France. It is not accidental, therefore, that Washington was to find it more difficult to penetrate here than, for example, in Morocco and Tunisia.

When Guinea alone of the French territories voted in a referendum on September 28, 1958 for political independence, France took economic, diplomatic and other forms of retaliatory action.

The US government, following in the wake of France, refused for several months to recognize the government of Sékou Touré. When such crude pressure proved unavailing, especially in view of assistance available from the Socialist states, Washington changed its strategy, beginning with such tactical initiatives as the appointment of an American Negro as ambassador, and later rolling out the red carpet for Touré on his visit to the United States in 1959.

Belgian colonialism, too, could not indefinitely remain indifferent to the tide of African national liberation, especially when it advanced to the French Congo, whose capital Brazzaville is only a short distance across the Congo River from Leopoldville (now Kinshasa). When the anti-colonial movement in the Belgian Congo, as reflected in strikes and political outbursts in early 1959 and a proliferation of "independence" parties during the following months, could no longer be contained, the Belgian government was finally compelled to begin negotiations which led to relinquishing its colonial rule on June 30, 1960. This was greeted by Washington with unconcealed satisfaction, soon to be followed by its own active policy.

In parallel with the anti-colonial armed struggle, an international political and diplomatic campaign was conducted by the newly liberated peoples and Socialist states to bring pressure to bear on the colonial powers to disgorge themselves of their African possessions. This took especially sharp form in the United Nations, where a loose Asia-African group began to function on an *ad hoc* basis in 1950 and assumed organizational form five years later. A separate African Caucusing Group formally took shape after the 1958 Conference of African States in Accra, specifically supporting the Provisional Algerian Government and campaigning for the setting up of specific dates for the independence of trust territories and colonies.

Although less involved in outright colonialism than its

NATO allies, the United States throughout the 1950's followed the timeworn colonial opposition to the establishment of timetables for the trust territories and colonies taken from Germany in World War I, or from Italy and Japan in World War II,[1] on the typical grounds that independence would be "premature". However, to improve the US image in the eyes of the anti-colonial forces, the US delegate more and more frequently abstained rather than vote against a resolution he had strongly opposed in debate. In 1957 the General Assembly finally adopted a Soviet resolution for steps to be taken to proclaim African trust territories independent within 3-5 years. Although severely modified, the resolution was approved, with the United States voting against.

In the general political and diplomatic imperialist opposition to African national-liberation forces, the United States played a major role in impeding and retarding the movement to independence. But the imperialist powers were less successful with the passage of time. In the UN Security Council, for example, they refused to permit discussion of colonial policies arguing that these were matters of "domestic jurisdiction" not within the competence of the international organization. However, since a simple majority vote can place an issue on the General Assembly agenda, the growing number of new sovereign states turned to this body with increasing frequency. But, on the question of Algeria not until 1960 were the anti-colonial forces strong enough in the General Assembly to pass—in the face of a French government boycott and Western-bloc opposition—a resolution favoring independence (63-8, with 27 abstentions). By that year, too, the 15th General Assembly by an overwhelming vote (78-0) called on South Africa, a colonial power in its own right, to end all racial discrimination in its trust territory of South West Africa; the United States and Britain opposed the resolution in debate and subsequently abstained. In the same year, the United States also opposed resolutions: to admonish Portugal concerning conditions of widespread forced labor, poverty, practically total illiteracy

[1] This applied to seven African (and four Pacific island) territories. They attained independence in the following order: British Togoland in 1957, French Cameroons and French Togoland in 1960, British Cameroons and Tanganyika in 1961, and Ruanda-Urundi in 1962.

and an oppressive regime in its non-governing overseas territories; to censure South Africa for its apartheid policies, and Belgium for its administered election in Ruanda-Urundi. US diplomacy generally could be found supporting, in actions, the position of its colonial allies, and urging conformance upon the subject peoples.

The close interconnection between economic, political and social questions showed itself in many ways during this period. Retarded African economic development, which had been accentuated by the postwar technological revolution, demonstrated that the industrially advanced capitalist countries were unable to satisfy the rising expectations of Africa as an appendage of Europe. A first step in resolving this paradox was sought by the African people in political independence. Afterwards, both sides were aware that the struggle for increased political and diplomatic strength, allies etc. would continue. This was reflected, for instance, in the determined position of the imperialist states in the UN regarding the formation of the Economic Commission for Africa. On the question of composition, the North Atlantic powers were so intent upon excluding the Soviet Union from this body that they were prepared, as a *quid pro quo* to forego US membership.[1] This led to their narrowing the base to exclusive African membership. On the question of including "social" development within the competence of the Commission, the United States and Britain, sensing that "social" implied broad implications which would threaten their monopolists' interests, led the opposition.

Perhaps most complex, and far from unraveled or measured, are US ideological forces and their influence. Their source and direction, even when not visibly germane to the African scene or issues, have had an important bearing on the pivotal question of furthering or hindering national liberation. During the transition period, for example, the postwar global political and economic ambitions of the United States were the source of its world-wide "anti-Communist" crusade, with such labels as "subversive" applied to disarm and fragment any militant opposition to imperialism. Not that this method was original—it was a century old and had flourished on the eve of the Second World War,

---

[1] See I. Wallerstein, op. cit., pp. 30-31.

particularly during the brief but violent quest of the Third Reich for world domination.

In Africa itself, US ruling class links, first to slavery and subsequently to colonialism, provided the economic and political motives requiring an ideological justification: thus, the European colonial powers provided "stability", independence was "premature", and subject peoples invariably required more years of education and preparation for self-rule—little of which, however, was actually forthcoming. In southern Africa, racist rule and apartheid, which found their parallel in the United States—although generally condemned and reforms urged—were attributed in self-justification by a white, Anglo-Saxon, Protestant American ruling establishment to a general "heritage of prejudice" and "human weakness", which would require gradual political and legal changes and education to overcome. American apologists urged that US policies had to take practical considerations into account and warned against the consequences of a war for independence, rather than against the intolerability of existing terror and brutality against overwhelming majorities.

In the transitional period, the achievement of African national liberation reverberated in government circles in a conflict between the supporters of "Old Europe" and the anticipators of a "New Africa", e.g., Chester Bowles, Adlai Stevenson, Wayne Fredericks, Mennen Williams, who advocated a new tactical approach. The evidence of progress in tropical Africa is ascribed by American educators, at best, to a "climate of anti-colonialism", and, at worst, to the "handing over" of independence. Thus, for example, Professor Rupert Emerson of Harvard University, in an effort at objectivity, writes that it is natural for Africans "to stress the heroic nature of their struggle" and for Europeans to "exaggerate their voluntary generosity in accepting African independence".[1]

In actual fact, the world-wide wave of national liberation promoted the transition for most of Africa from colonial rule to political independence in the late 1950's. Thus, at

---

[1] *Africa and the U.S. Policy,* Prentice-Hall, 1967, p. 4. Also, see Chester Bowles, *Africa's Challenge to America* (University of California, 1957) and Vernon McKay, *Africa and World Politics* (N.Y., 1963).

the beginning of the decade, there were only four sovereign countries on the continent, namely, Egypt, Ethiopia, Liberia and the Union of South Africa (the overwhelming majority of whose population, however, were subject to domestic subjection), which comprised peoples living in only 12% of Africa's area. By the end of 1959, there were already 10 independent countries, totalling 28% of the continent's area. But by the close of 1960, with another 17 new states appearing, 70% of Africa's total area inhabited by 75% of the continent's population consisted of independent countries. This could not fail to have its impact on America.

In the US government, a separate Bureau of African Affairs with its own Assistant Secretary of State was established in July 1958. This was involuntary recognition that Africa would no longer be regarded as an extension of Europe. It was to be viewed as an independent area, for which the United States would conduct its own aggressive policy.

In sum, during the transitional years of the late 1950's, although US political strategy, aimed at slowing up or halting the further advance of the national-liberation movement in Africa, showed a preference for the neocolonialist approach, which would conceal its conflict of interests with the already independent states, it had no intention of discontinuing its policy of support for the remaining minor—but extremely important—segment of colonialism in Africa. It feared that the anti-colonialist assault might not stop at half-way measures in settling accounts with the prevailing colonial regime, but its momentum might, in the process of liberation, go all the way towards ousting foreign monopoly capital and imperialist influence. And this not only from colonial Africa, but from an entire continent, which represented a major preserve of the world capitalist economy.

3. AMERICAN CIVIL RIGHTS MOVEMENT

The American people's struggle for Black Americans' civil rights, previously considered almost exclusively an internal question, received a particularly strong impulse from external forces in the 1950's. Thus, the anti-segregation Supreme Court decision of 1954, it is generally recognized, was influenced by the rising prestige of world Socialism, as well as Asian and African freedom struggles.[1] Similarly, the bus boycott in Montgomery, Alabama, in 1955, the affair of Little Rock in 1957 and the Freedom Movement.

By the close of 1960, with most of Africa independent and confronting the rest of the continent on the issues of colonialism and White supremacist rule, the link between Africa and the US scene became very close. The American Black, who had previously dissociated himself from his African background in large part because of its colonial identification with subjection, now began to identify with and to emulate African political and social struggle. Race discrimination in southern Africa as well as in the South of the United States, were subjected to world debate and action. In the UN arena, apartheid and race relations affected international relations.

Racial and social discrimination suffered in the United States by African diplomatic personnel, e.g., in housing, restaurants and transportation,[2] were thrown into glaring relief. Dramatic international incidents resulting from discrimination, such as refusal to serve African diplomats in restaurants, on the northern Route 40 between Washington, D.C. and New York City, developed. The indignation, protest and action of African diplomatic representatives and exchange students in the United States compelled the State Department to step in to enforce the norms of international behavior.

The second-rate-citizen status of the American Black led to some unexpected incidents in Washington's relations toward new Africa. During the 1950's, the US mission in Liberia had been staffed exclusively by American Negro technicians. This was interpreted by Liberians "as a special form of discrimination". It was not until 1959, however,

---

[1] "Most of the Negroes I know," wrote James Baldwin, "do not believe that this immense concession (the 1954 Supreme Court decision outlawing segregation in the schools) would ever have been made if it had not been for the competition of the Cold War, and the fact that Africa was clearly liberating itself and therefore had, for political reasons, to be wooed by the descendants of her former masters." *The Fire Next Time,* New York, 1963, p. 101.

[2] "Landlords will not rent to them; schools refuse their children; stores will not let them try on clothes; beaches bar their families," according to Edward R. Murrow, head of US Information Agency. *Herald Tribune,* May 25, 1961, International Edition, Paris. All further references to this source are to the International Edition.

that replacements began to be made on a non-racial basis.[1] Furthermore, when the State Department, seeking to curry favor, deliberately appointed a Black diplomatic representative to one new state, it was met with the rebuff: "Please send us your first-rate citizens."[2] This could not help making it plain to US policymakers that the actual status of the entire Black people in America was of basic concern to independent Africa, which would not be mollified by a few token gestures in the direction of racial equality.

In the sixties, greater recognition of the interconnection between the United States, Africa and the world led to the broadening and internationalization of the American civil rights movement.[3] Since 1960, wrote the chairman of the Student Nonviolent Coordinating Committee in 1964, our people are "conscious of things that happen in Cuba, in Latin America and in Africa".[4] In early 1966, at a UN luncheon tendered by the chief delegates of 15 African countries to the young civil rights worker Julian Bond, unseated by the Georgia State Legislature, Rev. Martin Luther King Jr., hailing the beginning of "a creative coalition between the black people of the U.S. and our black brothers in Africa", declared: "We have a kind of domestic colonialism in the U.S.—Harlem, Watts, the west and south side (of Chicago) ... we are determined on our side to cast off the yokes of our colonialism."[5]

A decade after the bus boycott in Montgomery, Alabama, to force desegregation of public transportation in the United States, Dr. King in 1965 was urging President Johnson to issue "unconditional and unambiguous" pleas for peace talks in Vietnam, and by 1967, he came out clearly against the Vietnam war (on the basis of its draining funds from domestic social programs). On April 4, 1968, this 39-year

---

[1] See J.D. Montgomery, *Aid to Africa: New Test for U.S. Policy.*

[2] Ibid.

[3] An early Communist analysis of theoretical aspects of the Negro question pointed to the strong ties with dark-skinned people liberating themselves throughout the world, and to Socialism which offers the fundamental solution. *Political Affairs,* January 1959.

[4] John Lewis, "A Trend Toward Aggressive Nonviolent Action", in *Negro Protest Thought in the Twentieth Century,* F.L. Broderick and A. Meier, the United States, 1965, p. 318. Moreover, as a result of the pioneering work of American progressives, the "masses and the Negro academic community feel a great deal of understanding and love for people like Robeson and DuBois" (loc, cit., p. 319).

[5] *The Worker,* January 25, 1966.

old proponent of massive direct action in pursuit of civil rights, a national leader who had become an international figure, was assassinated.

By the second half of the sixties, it was becoming a commonplace even for moderates and US government officials to draw the analogy between racism in South Africa and in the United States.[1] Their conclusions, however, led into harmless channels. Thus, whereas Ambassador Goldberg was defending the US position[2] of limited sanctions against Rhodesia and no action against South Africa, some American civil rights leaders, recognizing the link-up with US monopolies were seeking to join hands with Africa in common action. When the Committee Against Apartheid, of which A. Philip Randolph was chairman, called a boycott of the Chase Manhattan and First National City Banks, the labor leader declared to both banks: "You are accomplices of apartheid and, in the eyes of black people in South Africa and everywhere in the world, partners of oppressors."

Similarly, Washington's policy of accommodating to apartheid was condemned by the American Negro Leadership Conference on Africa, e.g., apropos of the visit of the USS F. D. Roosevelt (aircraft carrier) to South Africa in February 1967, apparently a move to reverse quietly a decision made a year previously not to use South Africa's deep water docks for naval purposes as long as segregation prevailed. With the beginning of open armed struggle on the part of the Zimbabwe people and African National Congress (ANC) against the racist regimes of Rhodesia and South Africa on August 13, 1967 American civil rights

---

[1] "The ghettoes of America," declared Ralph Bunche, "are like the native reserves in South Africa. They symbolize the Negro as unacceptable, inferior and therefore kept apart." The President's National Advisory Committee on Civil Disorders, in a report issued in March 1968, warned of "a kind of urban apartheid" coming to the United States with enforced Negro residence in segregated areas and semi-martial laws in many major cities. (See *The Economist,* March 9, 1968). A further presidential committee report, issued on July 17, 1968 by authors P.L. Hodge and P.M. Hauser of the University of Chicago Population Research Center, indicated that in Washington, D.C., the Negro population exceeded two-thirds of the city's total, while the suburbs were almost entirely white. The same trend was visible in other cities.

[2] *The Worker,* February 7, 1967. Roy Wilkins of the NAACP also urged support for the Johnson Administration on the grounds that it was under attack by Right-wing forces.

groups indicated their full support—up to Students Nonviolent Coordinating Committee's (SNCC) avowed intention of helping African guerrillas to fight (to prepare psychologically, it was reported, for a "Black International").[1]

The different emphases given in the triangular African, American and world force relationship have produced differing conceptual configurations and courses of action. These vary from the strong weight given to nationalism and race by Richard Wright, who advised Nkrumah to steer clear of the struggle between capitalism and Socialism, to George Padmore's concept of Pan-Africanism along Socialist lines but apart from broader world trends, to those like Frantz Fanon who have viewed Africa and the underdeveloped countries as an entity in itself even if "strengthened by the unconditional support of the Socialist countries.[2] Stokely Carmichael and other Americans in the civil rights struggle are prone to see the common relationship between the black people of Africa and US essentially in terms of racism,[3] as contrasted with James Foreman of SNCC who has emphasized that "... exploitation results both from class positions as well as race".[4] On this critical question some leaders of the Black Panther Party have taken a similar position.[5]

Initially, the Black Muslims, a movement which had its origins in protest against segregation in Christianity ("the white man's religion") and sought equality in Islam, produced a separatist nationalism with black capitalist tendencies.[6] The prominent leader Malcolm X, however, in searching to broaden the emphasis into the "extra-religious strug-

---

[1] *Herald Tribune*, August 30, 1967; also see A. Lerumo, "Our People in the U.S.A." in the *African Communist*, No. 33, 1968.

[2] F. Fanon, *Wretched of the Earth*, p. 62, cited in C.L. Lightfoot, *Ghetto Rebellion to Black Liberation*, N.Y. 1968, p. 128. See Chapter 12.

[3] This was accentuated to extremes after his three years as an expatriate in Guinea, when he declared in an interview: "We are not black Americans. We are Africans." In searching for a pan-African ideology, he contended that Americans with African ancestry should abandon the United States in a mass exodus—"Our primary objective should be Africa." *Herald Tribune*, June 16, 1972.

[4] *The Worker*, January 7, 1968.

[5] See, for example, Bobby Seale, *Seize the Time*, N.Y., 1970. (More recently he has embraced Black capitalism as a course of action.)

[6] This, despite its relatively strong working-class composition. See C. Eric Lincoln, *The Black Muslims in America*, Boston, 1961, Chapter I.

gle for human rights"[1] as well as internationally, especially after his trip to Africa (from Mecca to Ghana) in early 1964, enlarged his view of brotherhood to embrace "all races, all colors".[2] His increasingly class approach[3] and growing personal influence contributed to his break with Elijah Muhammed. He was assassinated on February 21, 1965.

The overemphasis on racism and nationalism, however understandable historically, has hampered American struggles for Black equality, and unfortunately, has spilled over to the international scene. Thus, the portrayal of the present world-wide struggle—of which the American civil rights movement is a part—as one being waged by the one-third white against the two-thirds non-white human population, or of the Christian versus the non-Christian world, has substituted skin color and religious belief for fundamental criteria for choosing allies.

From a world historical viewpoint, Marxism-Leninism understands class domination and exploitation, including its social aspects of racial discrimination, as both national and international. Since the socio-economic question of racism in southern Africa, moreover, falls within the framework of colonialism and is maintained with the help of world imperialism, the struggle and solution appear logically to be dependent upon a combined international effort of all anti-imperialist forces.

---

[1] *The Autobiography of Malcolm X,* 1964, (1966 ed.), p. 354.
[2] Op. cit., p. 362.
[3] Op. cit., pp. 365, 371, 416.

III. US NEOCOLONIALISM

1. GENERAL ELEMENTS AND POLITICAL FRAMEWORK

Why and when the United States went over to the new colonialism in Africa flow primarily from world and African objective conditions, including its own position, as well as from its policymakers' estimate of strength relationships and tendencies.

The latter part of the 1950's was marked globally by the growing all-round strength of the Socialist community (especially the Soviet Union), an international working class spearheaded by a united Communist movement, and the sweep of Asian and African national-liberation struggles. In Africa, national and social efforts fused into a unifying force of opposition to colonial rule and exploitation. The struggles in North Africa, in particular, by weakening a common enemy, helped to reinforce the independence movements in tropical Africa, which in turn exerted pressure on the rest of the continent. These, on the whole, centripetal progressive forces, confronting the traditionally rival colonial powers in their separate empires, with the United States playing a minor role, foretold a continuation of the progressive upswing.

This tendency was sensed also by world imperialism and especially by the United States with its global viewpoint and ambitions. Thus, a basic study issued by the US Senate in 1959 concluded that the dynamic character of the African peoples' drive toward self-government indicates that "the colonial system in Africa, as elsewhere, is fast running its course"[1] and that US policy "should be guided by the expectation of the primacy of Africans in all Subsaharan Africa".[2] Furthermore, in the face of the prevailing African

[1] US Senate, Committee on Foreign Relations, op. cit., p. 78.
[2] Ibid., p. 13.

militancy, the same study urged that hostile confrontation be avoided since "the more peaceful the transition to self-rule, the greater the likelihood that the present orientation to the West will be maintained".[1] This called for adapting to new conditions and methods if imperialism was not to be swept away by the storm.

On the basis of this flow and interpretation of events, US relations with the awakened black continent were geared to the general aim of keeping the new states in the political orbit of the "free" world, with predominantly capitalist economies, and under greater US influence.

Three distinct elements are discernible within US foreign policy. *First*, as in the previous period, the overall relations of the United States to its imperialist allies—its predominant position in the NATO political-military alliance aimed at bolstering world imperialism by continuing the cold war crusade against the Socialist community, national-liberation movement, and international working class. Secondary to this are sometimes very sharp differences between the imperialist powers, largely over relative politico-military position, as well as economic rivalry between them. *Secondly*, the relations of US imperialist allies to their colonies and ex-colonies—the form of struggle waged, and both old and new colonial methods employed. Differences of approach between the US and its partner/rivals sometimes have led to bitter conflict, e.g., in the Congo immediately after independence. *Thirdly*, and quite peculiar to the present period, the distinctive and growingly important political strategy of American imperialism to Africa as a continent, as well as its tactics in individual countries. These three main vectors, generally present and interacting in the form of pushes and pulls, account for the force and direction of present-day US imperialist policy.

To attain its objectives, US imperialism has sought to maximize its strength. In a primary effort to consolidate the overall imperialist position and harmonize rival interest in independent Africa, the United States and its NATO partners broke down the continent into "spheres of responsibility", roughly corresponding to previous spheres of influence or colonial rule. Although inter-imperialist contradictions as such may not have diminished, they have been

---

[1] US Senate, Committee on Foreign Relations, op. cit., p. 13.

overshadowed by the urgently felt need of the ruling classes of America and Europe to band together. This has involved a growing US emphasis on joint action, for example, in political-military blocs, and in multilateral financial institutions such as the World Bank and International Monetary Fund.

The second major imperialist pattern of strength relationships was a legacy of the disintegrating colonial empires. In the new sovereign states, centuries of colonialism had left myriad ties which bound them to the metropolitan countries in various degrees—less so countries which had been locked in violent struggle with their former overlords, and more so those which achieved self-rule relatively peacefully and where the colonial power had prepared the ground in advance. Apart from the independent role of the United States in such countries as Liberia, Ethiopia and North Africa, Washington's influence has been broadened keeping in mind its relationship to the ex-colonial power. Thus, where strong political, economic and military ties to the new states were retained by Britain, the United States found it advantageous to operate in conjunction with the former colonial power (in Libya, Nigeria, Ghana and Sierra Leone). This was less true of the more competitive US and French relations, for example, in North Africa. Where the individual ex-colonial power was not so strong in relation to the national-liberation movement, and unable or unwilling to develop the necessary sophisticated forms of domination in cooperation with the United States, e.g., Belgium in the Congo, American imperialism came into hostile confrontation with it and found greater opportunity to conduct its own aggressive strategy.

In the three-fourths of the continent in which the colonialists had to relinquish state power, naked political and military rule had to be abandoned. Nevertheless, subordinating relationships and disparity in strength continued in various spheres.

These unequal relationships were utilized by imperialism as new colonial weapons. Thus, for example, the political ties of the NATO powers were made use of to influence the diplomatic relations of their ex-colonies in the United Nations. Investment, aid and trade were used not only as economic levers to derive high profits but also with political strings attached. Economic advantage and social status were employed to promote certain social strata and individuals. Language, education, religious and cultural ties, frequently buttressed in the form of technical assistance, took on increasing importance as means of exerting broader ideological influence. Military bases and agreements, which encroached upon sovereignty, served as pressure points up to and including intervention. Although many of these weapons were not new, a greater variety developed and were perhaps more widely employed than in previous periods.

The general application of such forms, methods and techniques under the new conditions of political sovereignty has come to be known as neocolonialism. This presupposed fresh efforts in all spheres on the part of imperialism, in general, and the United States in particular. The latter, which traditionally leaned on indirect and veiled domination in Latin America and the Far East, became an early advocate of a similar approach in independent Africa. This may not be unrelated to the fact that the United States enjoyed preeminance in the economic field, which it was anticipated would in the future predominate, and the lack of identification of Washington with colonial rule gave it an initial political and ideological advantage over its partner/rivals.

The common denominator of neocolonialist policy, nevertheless, is not the form or method—all methods are employed—but the aims and objectives pursued in the new period. The special emphasis on economic instruments reflects a recognition of the particular vulnerability of the emergent states in this vital sphere, their need for funds, equipment and technological know-how to overcome centuries of economic backwardness. Despite the importance of this sphere, however, it would be an exaggeration to equate economic weapons with the new colonialism as was sometimes done in the early sixties. For, in addition, many other difficulties of emerging nationhood are exploited. In the social sphere, for example, the predominance of communal and tribal organization, which is reflected in a lack of national homogeneity and consciousness; the embryonic stage of development of the working class and trade union movement, which imparts unusual importance to individuals and élite groups from other social strata; a lack of trained and qualified personnel in most civilian fields, which makes technical assistance such an attractive avenue of influence. Particularly critical in tropical Africa, for instance, has been the absence of a nationally trained officer corps and a strong military establishment, which has opened the gates to inspired coup d'états and foreign intervention. The US employment of levers—both governmental and private—in various spheres has been flexible and pragmatic, shifting in emphasis in various countries and periods. For analysis purposes, some of these may be more conveniently viewed by sphere, although in practice, to be sure, they are closely intertwined.

* * *

The imperialist powers in the colonial period were able to pursue their economic or strategic aims in Africa mainly through the medium of state power—the political control of a bureaucracy, the armed forces and state apparatus. Conversely, political independence was generally recognized by the national-liberation movement as a precondition for economic and social progress. For both sides, moreover, the primacy of politics, recognized or not, has continued in the post-colonial period.

Present-day US foreign policy in Africa is essentially *political*—class and national, and not narrowly a consequence of local economic interests. Furthermore, the rulers of America frequently promote economic interests and employ levers in all spheres to further their continental political strategy. In general, this is part of their overall aim of preserving world imperialism, which is based *in the long run* on the profit system in which US dollar considerations predominate.

As contrasted with US imperialist political aims and economic interests, which are generally meshed, independent Africa reveals wide discrepancies between political and economic factors. Thus, an urgent general problem facing all the new states of Africa is the glaring contrast between their political independence and economic dependence. They are seeking, in various ways and degrees, to resolve this lag through the achievement of greater economic independence (in conjunction with appropriate domestic processes). US foreign policy, on the whole, is striving to resolve this discrepancy in the opposite direction, i.e., to make use of the economic vulnerability and other weaknesses of the newly emerging states in order to tie them politically to imperialism. These two opposing sets of aims constitute a basic and continuing antithesis, which cannot be ignored or smudged over without losing much of the dynamics at work.[1]

Although late in the Eisenhower Administration, but particularly under President Kennedy, the United States expanded its economic activities, nevertheless, the major emphasis of its foreign policy was and has continued to be *politico-military*. This is the tendency, or bias, of the cold war. To escalate it is to go to hot war—an attempt to impose a political decision by recourse to force. This took place, for example, in the complicated US political maneuvers of the UN force in the Congo following independence, and more clearly during the US-Belgian-British military intervention in the Congo in November 1964. To de-escalate it would be to accept economic competition as the main arena of struggle between different social systems, which, in conjunction with a policy of non-intervention by outside powers, could provide preconditions for a more peaceful continent. This concept of peaceful coexistence, however, is opposed by those US extremist forces which are obsessed with military escalation as a mechanism for changing the balance of forces. Escalation from the political to the military sphere, and further refinements into a game theory à la Herman Kahn, however, are in essence a "big stick" policy, which instigates local wars and increases the danger of nuclear world war, rather than resolving the problems at issue.

Thus, it was no less a figure than the late President Kennedy, for example, who clearly reminded West Point graduates that: "The basic problems facing the world today are not susceptible of a final military solution . . . neither our strategy, nor our psychology as a nation—and certainly not our economy—must become permanently dependent on

---

[1] Officially this was implicit in the form of a dilemma presented by W. W. Rostow, formerly Chairman of the Department of State Policy Planning Council: ". . . the United States finds itself often in a rather complicated position. Our friends in the developing countries are, in one part of their minds, pleased to receive our help and support; but, in another part of their minds, one of the major purposes of revolutions of nationalism and modernization is to achieve a higher degree of independence of the more advanced powers of the world and in particular a higher degree of independence of the US." (W. W. Rostow, "U.S. Policy in a Changing World" in the *Department of State Bulletin,* November 2, 1964, p. 642.)

an ever increasing military establishment."[1] The logic of this would imply the pursuit of a politico-economic policy of peaceful coexistence.

But what *is* the actual relationship between *force* and *politics* in the conduct of US foreign policy?

Although US objectives "are in the largest sense political, not military", wrote the former State Department's chief policy planner, "our basic national security policy now accepts the central reality of ... controlled, limited, politico-military confrontation. These episodes are not regarded as exceptions to the rule, to be dealt with *ad hoc*, but as the form the struggle is most likely to assume."[2]

In Africa, too, it has been military power and its actual, or threat of, employment which have been considered the background force necessary to achieve US political objectives. It is important to note that this officially enunciated general viewpoint has been applied—even if not as directly as in Vietnam or Latin America—and tested in practice in Africa. Of no less importance is that such a military force theory is a fallacious concept—being one-sided—and is proving so in fact: US military capabilities frequently cannot be brought to bear, are inexpedient or insufficient to constitute "the inescapable backdrop for the whole of (its) civil policy".[3] Moreover, although individual military victories in themselves may temporarily hold back progress through repression, they cannot solve political, economic and social problems, which again clamor for solution.

In the author's treatment of the interrelated elements of American imperialist strength and weakness in each sphere —from economic to military—it will be noted that politics runs like a red thread throughout, even hyphenated to political-economic and political-military, rather than dealt with separately. This has been found most realistic and the most useful organizational form of presentation—although it is readily acknowledged that others may find models or forms of analysis which may be no less fruitful or practicable.

---

[1] *The New York Times,* June 7, 1962.

[2] W. W. Rostow, *View from the Seventh Floor,* N.Y., 1964, p. 34.

[3] Ibid. To various degrees, this is recognized by the more perceptive US foreign policymakers, e.g., Senator J.W. Fulbright in *The Arrogance of Power,* 1967; former Secretary of Defense R. McNamara in *The Essence of Security, Reflections in Office,* N.Y., 1968; Henry Kissinger in *Agenda for the Nation,* Brookings Institution, Washington, 1968.

2. ECONOMIC BASIS

INTERRELATIONSHIPS

US economic relations with Africa have been molded essentially by the colonial legacy and the post-independence general pattern of imperialist relations. The latter, in turn, have been mainly an outgrowth and continuation of the exploitation of natural and human resources, which has produced backward, one-sided economies geared to and largely dependent on the export of primary products to the capitalist powers. Thus, 90% of Africa's exports (1967) comprised minerals and agricultural products: oil, unrefined copper, and coffee constituted 37%; cocoa and cotton—13%.[1] The picture is even more lopsided in tropical Africa, where about five-sixths of the output of 33 countries consists of 7 agricultural crops.[2] Not that the continent lacks a balance of resources. It is rich in land, minerals and hydropower. However, the mineral potential, which is considerable but largely in the hands of foreign monopolies, is drained out to their industries abroad at great profit. And low-productive agriculture, in which some 77% of the population are occupied, is still either on a subsistence basis or engaged in what has proved to be a futile effort to increase foreign exchange by expanding export crops to the ex-metropole. (See: Trade section.) This has been a major trend rather than that of advancing together with industrial development into more balanced and integrated economies.

Largely as a result of past and present imperialist policy, a continent comprising some 310 millions, or about 8% of the world's population in 1965, contributed only 2% of world output although about 5% of its exports. A per capita annual rate of growth of less than one percent in the past century[3] has hardly improved since independence. Net domestic product rose 1% in 1960-64 (instead of 5% as envisaged by the United Nations).[4] In the period 1960-66, for 80% of the continent's population there was an annual rate

[1] Third Report of the Economic Commission for Africa, *Economic Conditions in Africa in Recent Years,* the United Nations, 1969.

[2] Cocoa (23%), coffee (19%), cotton (16%), groundnuts (14%), tobacco, rubber and sisal together=14%). William Allen, *The African Husbandman,* London, 1965, pp. 461-64.

[3] Third Report of the ECA, op. cit.

[4] According to UN figures, the GNP increased by only 3.7% compared with a 2.7% population increase. See *North Africa,* September-October 1966.

of growth ranging from 0 to 2% instead of the minimal 3% which the United Nations estimated was necessary in the "Decade of Development" to satisfy the aspirations of the people.[1] Consequently, for four-fifths of the African people the annual per capita income was still less than $100. On a world scale, the average annual income per capita in industrially advanced countries rose from 1960 to 1965 about $220 to an average of approximately $1800, while in the underdeveloped it rose by about $7 to $90.[2] Thus, the actual income gap between the developing and advanced industrial countries has continued to widen, with the annual increase in production in the United States alone, for example, equal to the total output of Africa.

The role played by imperialism in exploiting and retarding the development of Africa during the colonial period is generally acknowledged and today can count upon few apologists. However, its present negative economic influence is far from being admitted or erased. Nevertheless, imperialism is being increasingly recognized as responsible for the lack or slowness of development in the new states. It has become a commonplace, therefore, for all political shades of African leadership to pose such questions as: Where is the aid, the softer loans, the reduced trade barriers, the effective commodity agreements, and the improved terms of trade?[3] In fact, by minimum UN standards, aid has been insufficient, credit terms have become harsher, and terms of trade have worsened.

None of these economic spheres taken separately will provide a satisfying answer to the why and wherefore of the economic picture. It is the general pattern of and interconnection between the various categories which is revealing—both for the individual country and for the continent as a whole. Thus, in southern and central Africa, if US (in conjunction with British) profitable investment is our point of departure,[4] then aid to infrastructure is linked to such

[1] Third Report of the ECA, op. cit.

[2] UN figures. See *Africa Report,* March 1967.

[3] "An Escape from Stagnation" by Tom Mboya, in *Africa Report,* March 1967, p. 14.

[4] A view frequently encountered is that the export of capital today does not have the same importance as it did in the early part of the century. Trade, for example, is given pride of place by some economists. See, for instance, *Le Pillage du Tiers Monde,* Paris, 1965, Pierre Jalée, pp. 95 and following.

investment, and trade flows in parallel. Export of capital has not proceeded evenly. If investment in the independent states tapered off in the late 50's and early 60's, this represented a short-term or regional phenomenon based on a "lack of confidence" in the safeness or profitability of investment, which changed in the second half of the decade.

Of cardinal importance is the political framework and direction of imperialist economic relations, especially so since political independence was not at all complete in most cases—largely as a result of the economic lag—and the struggle is still going on for its realization in both spheres. This would imply the necessity for a political-economic, rather than a strictly economic analysis, or an economic analysis with the political kept in mind. From the standpoint of the importance of categories in the economic sphere itself, i.e., magnitude, stability and profits, US private investment is primary. This is followed in our study by aid and trade for the logic of the general economic pattern.

Although not as large in absolute terms as in other continents, US investment, aid and trade in Africa is of particularly high economic and political leverage in a continent which, except for South Africa, is economically underdeveloped. In the present decade, these have increased (although not uniformly) more rapidly in amount—and in importance—than during any comparable previous period.

In the *first* place, under colonial rule, *political control* permitted the metropole to grant its own monopolists a privileged position and economic penetration by others was hindered, e.g., by restrictions on their investment and the transfer of profits, and by disadvantages in tax payments. After the achievement of national sovereignty, however, the loosening of political ties to the former colonial power permitted significant US economic inroads.

Secondly, US imperialism attaches great political importance to the governmental policy of generally shoring up imperialism and, at the same time, assisting American monopolies to penetrate the newly independent states in search of high profits. In 1960, for example, when foreign capital tended to flow out of the new states, the US government sought to reverse this flow: a 30% tax credit was granted for new investment; risk guarantees were increased to cover investments; investment survey costs were shared by the government. US policymakers have shown particular interest in the political implications of an African capitalist economy. To strengthen an African state's private sector, for instance, US loans have been extended to the country's industrial development banks which, in turn, make loans to local businessmen. It is openly aimed, thereby, to foster a local bourgeoisie which will have a stake not only in co-operating with American monopoly capital domestically, but also in matters of foreign policy. More than ever before, the dollar, both private and public, is viewed by Washington as a political dollar.

The economic and political ties of the imperialist powers are interwoven in various major blocs, vertically linking them with their ex-colonies, e.g., the early transmutation of the British Empire to Commonwealth, and the later formation of the French Community in 1958. In the roughly parallel monetary systems, 31 of the 39 independent states in 1967 were in either the sterling area (14) or in the closer knit franc zone (17). The much smaller US sphere included Liberia, which is in currency union with the United States and uses the dollar internally; and Ethiopia (as well as French Somaliland, largely because of its close trade ties with Ethiopia) is also frequently considered in the dollar area, although not formally so.[1] With the exception of those countries which are not in any such blocs, (e.g., ARE, the Sudan, Algeria, Guinea) the other African countries are members of the Portuguese peseta or South African monetary areas (the latter very closely linked with London).

The "Six" members of the European Economic Community formed a new major imperialist bloc "Eurafrica" in 1958 which drew into it as associate members 18 African states—17 former French colonies and the Congo (Kinshasa), now Zaïre[2]. This gave the monopolies of the non-colonial powers, particularly West Germany's Krupp, Mannessman and Thyssen, an opportunity to penetrate Africa on equal

---

[1] See A. Kamarck, *The Economics of African Development,* Praeger, 1967.

[2] In adopting the name Republic of Zaïre on October 27, 1971, in lieu of the Democratic Republic of the Congo (the official designation since 1967), the Kinshasa government was reverting to the original name of the Congo River, "Zadi" (big water), told by the people living on its banks to early Portuguese explorers in 1484. This became Zaïre in Portuguese, applying to both the river and country.

terms with France's Lazard, Rothschild and Pechiney. For the African states, the attraction was to be secure markets at higher than world prices and a Development Fund of $200 million for the first 5 years 1958-62 (only a fraction of which was made available, mainly for infrastructure). But, under pressure of the FRG and the Netherlands, the renegotiated 5-year agreement at the Yaounde Convention provided that goods would be sold at world prices by the end of the period 1964-69, thereby depriving the young states of compensating "subsidies for production".[1] A decade of experience in this bloc has confirmed that the principles of customs union, free exchange and free movement of capital prevent or retard industrialization in countries too young to develop infant industries without protection against the more powerful European monopolies.

The further widening of this bloc in July 1966 to include Nigeria, which did not thereby give up her trade flow and membership in the Commonwealth (and in turn received only limited free entry for her most important exports and no financial aid) paralleled the inconclusive negotiations of Britain for membership in the Common Market.

US general support for European integration, which was based on broad political, military and economic considerations, had its counterpart in qualified US support for "Eurafrica". Washington's major objection was to the EEC's policy of support prices and preferential arrangements ostensibly for economic—but also for political—purposes, which reflected the French "closed-system" of trade. The US-advocated "open-system" of trade,[2] it was hoped, would give Washington greater access to and influence in the French-dominated bloc in Africa.[3]

The failure of "vertical" integration with the European powers to overcome economic fragmentation and dependence

---

[1] See *West Africa,* March 25, 1967; and *Comment,* June 3, 1967.

[2] Both Britain and France, in general, have sought to broaden their preferential system from the colonial period and to tie it to the EEC, while the United States, with no extensive colonial legacy, usually is opposed. See, for example, the interview with George Ball, then Deputy Under Secretary of State, in *Realité,* July-August 1968.

[3] Although sharp rivalries continue between the European powers (e.g., the FRG and the Netherlands complain that French firms get the lion's share of contracts for projects financed by the Development Fund; pressure is put on Italy to contribute more to the Fund), the

on a few cash crops, to mobilize greater resources for development and to broaden markets, increasingly inclined the new African states toward "horizontal" African cooperation. Early examples of continental joint action were the establishment of the Organization of African Unity, the Economic Commission for Africa and the African Development Bank. In the second half of the sixties, the young states looked increasingly to regional cooperation as a desirable form of economic organization.

Regionalism was also of particular interest to the United States in that it either opened up or cut across rival blocs and provided greater possibilities for penetration. To ensure that the new states did not pull away from imperialism Washington has sought to encourage and actively participate in regional projects[1] to coordinate its direct and indirect influence in both old and new regional groupings.

In East Africa, for example, the United States helped give a new impulse to the economic union of Kenya, Tanganyika and Uganda (inherited from the British colonial East African Common Services Organization of 1927), which was gradually disintegrating by the beginning of 1966.[2] Edward M. Korry, US Ambassador to Ethiopia, was especially active in encouraging regional activities. In May 1966, for example, a provisional treaty of association was signed by an East African group of 7 countries—Ethiopia, Kenya, Tanzania, Zambia, Malawi, Burundi and Mauritius. At the same time, the US government sought to direct its activities regionally. Thus, Washington indicated that it would grant aid in future on a regional basis. A similar policy was initiated by the World Bank, and American cor-

---

EEC "agree in a common approach" before negotiating with the African states for renewal of the agreement. *The Economist,* April 27, 1968.

[1] See President Johnson's first address on Africa on May 26, 1966, in *The Department of State Bulletin,* June 13, 1966.

[2] The East African common market had been viewed as a mixed blessing biased toward non-African interests through the favored excolony, so that about 60% of trade and all manufacturing industry was concentrated in Kenya. To hold the market together, the Kampala agreement of May 1964 had allocated certain industries to Tanzania. See *African Diplomacy, Studies in the Determinants of Foreign Policy,* ed. by V. McKay, London, 1966, p. 65, and "The Integration of Developing Countries, Some Thoughts on East Africa and Central America" by Aaron Segal, in the *Journal of Common Market Studies,* Vol. V., No. 3, March 1967.

porations, too, followed the governmental lead, especially in the vital field of communications. In central Africa, the Congo became the pivot of an initial group (including Chad and the Central African Republic), which appeared interested in expanding and cutting across the central African Economic Union of 14 French-speaking states.

Foreign control is still proving to be a key obstacle to economic cooperation. In West Africa, the agreement concluded between Guinea, Liberia, the Ivory Coast and Sierra Leone foundered as a result of political differences, accentuated by blocs to which they belong.[1] Thus, in currency and banking, while Guinea has its own franc, Liberia is in the US dollar area, Sierra Leone in the British sterling zone and the Ivory Coast in the French franc zone. In a new group of 12 West African states, which signed an agreement in Accra in May 1967 looking to the creation of a common market of West Africa (Guinea was not included), political problems made themselves felt with respect to customs and industrial development.[2] African interests which need to be coordinated become complicated and overshadowed by foreign monopoly interests. Thus, the building of a heavy industry,[3] as envisaged by the UN Economic Commission for Africa, is to be financed and dominated by foreign capital. This, by its very nature, involves political questions, which cannot be avoided by urging—as does Washington—that the new states should concentrate on economic measures and not be diverted by political questions.

INVESTMENT, PROFITS AND US POLICY

GENERAL

The underlying economic basis for US capital export, which has far from lost its significance even if paralleled and overshadowed by political considerations (examples of

---

[1] See *West Africa*, May 22, 1967.

[2] Ibid.

[3] By 1980, according to Robert Gardiner, Secretary of the ECA, a force of 500,000 workers would be employed, providing 75% of the region's industrial production. See *Мировая экономика и международные отношения (World Economy and International Relations)*, April 1967, p. 124.

interrelationships will be dealt with later), resolves about ensuring the supply of certain minerals and raw materials and their profitability. US monopolies, as might be expected, sometimes play down their needs for primary products since it affects their bargaining position vis-à-vis the developing countries, and point to the declining requirements of modern industry for a few natural products, e.g., rubber and fibers, because of substitutes (or to food grain where the United States is a net exporter), or to a relatively favorable domestic position fulfilling most of their requirements, or to alternate sources either geographically or technologically. Some of these are either half- or abstract truths which serve chiefly to hide certain hard practical realities.

On balance, in fact, the present high US gross national product essentially derives from manufacturing (about 30%) which depends greatly on imports—the cheaper, the more profitable—of raw materials, e.g., for all of the major metals except iron, more than half of American industrial needs come from foreign sources. Furthermore, despite the fact that the United States is the world's largest producer of oil, it imports about 20% of requirements and authoritatively is described as "irrevocably" a net importer,[1] with prospects of constantly increasing future needs for energy and petrochemicals, both domestically and by affiliates abroad. Although Canada and Latin America are its dominant sources of mineral supply, Asia and Africa are nonetheless extremely important and the most probable sources for supplying future oil demands.

US monopolies, not surprisingly, have shown a prime interest in the continent which provides more than one-half of the capitalist world's mineral exports (1968): leading in gold, diamonds, cobalt and chrome, and important in manganese, copper, vanadium, uranium and asbestos. The United States, for its part, imported from Africa in the same year a significant proportion of its iron and ferroalloy ores—manganese 56%, chromite 39%, cobalt 27%, and iron ore 7%; non-ferrous metals—antimony 29%, copper 9%; and also, among other commodities, rubber 15%, fibers 10%, oil 9%.

---

[1] See *An Appraisal of the Petroleum Industry of the United States*, US Government Printing Office, Washington, D.C., 1965; Donald J. Patton, *The United States and World Resources*, N.Y., 1968.

American monopolies' specific aims of ensuring a rich supply of cheap minerals and metals, oil and gas, and tropical products have been more than amply fulfilled, as well as the more general financial aims of realizing high profits on the export of capital (as will be shown). Indicative is the fact that US monopolies have been engaged mainly in the continent's richest mineral and oil regions—the gold of South Africa's Witwatersrand, the oil deposits of the Afro-Arab countries, the Copperbelt and Katanga "mineral scandal", and the bauxite, manganese and iron ore of West Africa. The oft repeated claims of US private enterprise, on the other hand, that the compensating concomitant of its investment is to help the Africans in their economic development, much less political independence, do not withstand scrutiny.

This judgement is borne out, in the first place, by the continued emphasis on mineral exploitation (extraction and export) for US capital, supplemented by a more recent affinity for oil. During the 10-year period (1958-67), for example, the biggest investment increase absolutely and relatively was in petroleum, which rose from 36% to 54% of the total, with Libya showing a more than 18-fold expansion. Moreover, oil together with mining, which included about 17% of the total, comprised 72% of US private investment as compared with 64% a decade earlier. Thus, the combined extractive industries can hardly be described as losing emphasis even in the "development decade".

US investment in manufacturing did increase during 1960-69 from $118 million, or 11% of its overall total, to $454 million, or 15%—a definite, even if slow percentage rise in view of the low level of industrialization in most of Africa. However, closer examination shows that this increase of $336 million was confined mainly to South Africa—$226 million as compared with merely $70 million going to the rest of Africa combined. Thus, at the close of the decade, US capital in the manufacturing industry of the racist state constituted $374 million or 50% (a rise of 12 percentage points) of American monopolies' direct investment there, as compared to only $80 million or 4% (a rise of 3 percentage points) in the rest of Africa. This markedly increased the former's industrial and technical strength both absolutely and relatively with respect to the latter, with attendant political implications.

Although US private capital in Africa, traditionally small,[1] is still not as large as that of the ex-colonial powers,[2] and in 1957 constituted 2.2% of total US investments abroad, the recent rate of expansion has been greater on this continent than in most other areas. Thus, book value of US direct private investment, which purportedly represents roughly half of total US (direct plus indirect) private investment, rose from $664 million at the close of 1957, i.e., the year the first black African state, Ghana, achieved independence, to $2.3 billion[3] at the 1967 yearend, or 3.8% of foreign investment (see Table I). This is an increase of al-

Table I

US Direct Private Investment*
(Cumulative book value at yearend, in million dollars)

	1957	1964	1967	1970
Total all areas	25,394	44,386	59,267	78,090
Africa, total	664	1,769	2,277	3,476
of which:				
South Africa	301	467	667	864
Rhodesia and Nyasaland	59	83	—	—
Libya	24	402	456	1,009
Liberia	72	189	173	201
Other countries	208	628	982	1,404
Middle East	1,138	1,332	1,748	1,645

* Sources: *Survey of Current Business*, August 1963; October 1968 and 1971.

[1] By 1936, private investment, according to S. H. Frankel, amounted to $6 billion from Europe, with the exception of about $100 million from the United States.

[2] Foreign capital in Africa was estimated at about $20 billion in 1963; Britain—$7.5 billion, France—$6.0 billion, Belgium—$3.5 billion, the United States—$3.0 billion, the FRG—$0.2 billion. *Африка. Энциклопедический справочник (Africa. A Reference Book)* in two volumes, ed. by I. L. Potekhin, M., 1963, Vol. 2, p. 68. The US position is now stronger—absolutely and relatively—especially if one takes into account the flow of funds from international financial institutions, much of which is from the United States.

[3] *Survey of Current Business*, US Department of Commerce, Washington, October 1968, p. 24. The market value is frequently about 2-3 times this figure.

most 3.5 times, as compared to a less than doubling of total US foreign investment in the world as a whole. By 1970, the share in Africa had risen to 4.4%. Although the framework for this, as previously shown, was largely political, we shall concern ourselves here with the economic motives.

The distribution of investment generally parallels not development needs but the location of mineral resources, which is geographically not even. Roughly, southern and central Africa are the areas of greatest concentration, with South Africa alone showing 25% of the total by 1970. In North Africa, Libya accounted for 26% by 1970. In West Africa, Liberia—6%. Hidden in official statistics in the meaningless and growing category "other countries" was 43% in 1967 (as compared with 30% a decade earlier) of the total, which is of substantial significance especially for concealing the country pattern of most recent US penetration (see later).

Drawn to the combination of rich minerals and cheap labor, possibly up to one-half of US private investment is in the largely *colonial* and *racist-dominated* intertwined economies of southern and central Africa, where it is linked especially with the much greater British and the predominant South African capital. For a long time, South Africa's gold and diamonds have constituted the continent's major minerals, and although many others have come to be exploited, gold continues to be the major metal of Africa. It is important to the United States both as a commodity and in relation to its monetary system. South Africa's rising production of gold reached about 65% of the capitalist world output in 1966 and was valued at about $1 billion.[1] This dwarfs the approximately $100 million produced by Rhodesia, Ghana and Zaïre taken together. Of lesser significance, but not unimportant, is that Africa (South Africa, South West Africa [Namibia], Zaïre, Sierra Leone, Ghana and Angola) continue to produce almost all of the capitalist world's diamonds, with the South African de Beers Co. controlling the international selling monopoly, and the US an investor, but more important—the main world purchaser.

---

[1] See A. Kamarck, *The Economics of African Development*, N.Y., 1967, pp. 139-40.

In this same southern and central political-economic complex, non-ferrous metals also have been an important target of US monopoly investment, and particularly important in the age of communications—the copper of Zambia with an output of about 800,000 tons per year. It is second only to that of the United States in the capitalist world, and could have represented either a big competitor, or a profitable supplement. For American monopolies it has become the latter. Zaïre is also a major producer, with an output of about 300,000 tons per year. Two-thirds of the capitalist world's cobalt, used for missiles, jet engines and motors, comes from Zaïre (85% of this; plus 50% of the tin, and 40% of the zinc), Zambia and Morocco, the United States importing about 90% of African production, mainly from Zaïre; two-fifths of the capitalist world's manganese from Ghana, South Africa, Zaïre and Gabon; and one-fourth of the uranium (Zaïre, Gabon, South Africa). South Africa and Rhodesia supply half the world's chrome ore (used for stainless steel, jet engines, armour and ammunition), one-third of its vanadium ore, and one-fifth of its asbestos. Thus, the fact that about three-fourths of Africa's mineral resources originate in South Africa, Zambia, Rhodesia and Zaïre has been the primary attraction of US capital in these states.

In the *independent states* of North and West Africa, US monopoly capital has flowed overwhelmingly to the extractive industries during the 1960's: the *oil fields* in Libya, Nigeria and less so Algeria; the *iron ore* mines in Liberia, Mauritania and Gabon; the *bauxite* deposits in Guinea, Ghana and Sierra Leone. This was prompted in the postwar period in no small part by a US economy using, on the one hand, ever increasing amounts of the world's raw materials and, on the other, its own reserves growing scarcer (as compared with the prewar period when it was considered relatively self-sufficient). Although substitution of resources is frequently feasible, through the more thorough use of those available, the development of new technology etc., to prevent any complete reliance on particular foreign sources, cost must be reckoned with constantly.

Thus, the distinctly financial attraction for the export of US capital in general revolves about gaining super-profits. However, if the magnitude of investment is difficult to determine, so much more so is the rate of profit.

The approximately 500 American companies[1] in Africa (of the 3,500-4,000 with direct investments abroad) may boast of the "contribution" of private capital to the developing states but are curiously reluctant to show the extent of capital export, nor, for obvious reasons, their profits. For this would reveal the high degree of exploitation in the extractive industries, where a relatively small investment and large cheap labor force make for a high rate of profit.

With no pretense of an overall analysis of US profits in Africa, one may note that certain highlights are visible even from official figures (Table II). Thus, after a relatively modest rate of profit in the initial years of the decade, direct private investment earnings, which are defined as distributed plus undistributed profits (before US taxes), averaged 18% in the 5-year period 1963-67, increasing to 21% in 1968 and 24% in 1970. Throughout the decade, a remarkably steady 20% profit rate from South Africa provides the bed-

Table II

Annual Earnings on Direct Private Investment* (book value)
(millions of dollars)

	1960	1964	1967	1970
Total, all areas	3,566	5,061	6,017	8,733
Africa, total	33	380	453	845
% profit	3%	21%	19%	24%
South Africa	50	87	128	141
% profit	17%	20%	20%	17%
Libya	a	258	292	557
% profit		68%	65%	55%
Liberia	b	18	16	28
% profit		10%	9%	14%
Other countries	—	—17	17	119
Middle East	734	813	1,004	1,176
% profit	64%	61%	57%	71%

* Sources: *Survey of Current Business*, August 1961, 1963 and 1964; September 1965, 1966 and 1967; October 1968, 1969, 1970 and 1971.
a — included in North Africa (—69).
b — included in West Africa (37).

[1] See *Africa Report*, No. 1, 1969.

rock of American monopoly profits and has more than economic implications.

However, the most fabulous profits came from petroleum, especially since 1964. Thus, Libya's oil alone showed a 66% rate of profit in the following four years, or about 3.5 times the overall average and represented over two-thirds of the total earnings from the entire continent.[1] (The equally large oil investments in other African countries largely show paper losses since they are in initial stages of development and exploitation.)

Moreover, North African oil should be viewed as a complex in conjunction with the gigantic US investment in nearby Middle East oil (equal to three-quarters of the US total in Africa), which showed a comparable rate of profit. These combined two oil regions, in which less than 4% of US foreign capital was nestled, brought American oil monopolies 22% of the total US overseas earnings in 1967. It helps to explain much of US policy both before and after the Israeli blitzkrieg.

But earnings (distributed plus undistributed profits) are far from a complete picture of profits as a whole since they omit inflated depreciation charges and depletion allowances and other sophisticated bookkeeping devices which constitute the art of profit-hiding and tax avoidance. The AFL-CIO, therefore, regards cash flow, i.e., profits plus depreciation allowance, as the "accurate measure of a company's return, since it is the amount of money left over after the payment of all costs and taxes".[2] Along similar lines, a number of American economists[3] realistically measure profitability by cash flow including changes in the price of stock. This reflects some of the company's hidden profits in the form of appreciation or growth in the market price of shares. It may

---

[1] Rate of profit (on book value) increased to 75% in 1968 and 80% in 1969, partly due to "disinvestment resulting from repatriating earnings in excess of current earnings". *Survey of Current Business*, October 1970, pp. 30-31. (Thus, profits from this country rose to three-fourths of the total profits from the continent.)

[2] *American Federationist*, June 1962, cited in *Profits in the Modern Economy*, ed. by H. W. Stevenson and J. R. Nelson, University of Minnesota, 1967, p. 35.

[3] Ibid. See, for example, the essays by Joel Segall of the University of Chicago, and D. Bodenhorn, Ohio State University. A lucid Marxist study on this theme is *The Income 'Revolution'*, by Victor Perlo, N. Y., 1954, pp. 42-44.

show up, for instance, when a controlling interest in a company comes up for sale, takeover, or nationalization. Moreover, this still does not include the funnelling off from the companies of funds not listed as profits into the personal hands of those who have a controlling interest through inflated management/owner salaries, fees, expense accounts, pension schemes, bonus stock and many other devices which have contributed to the building of multi-millionaire fortunes.

An inkling of how much difference may be involved between paying dividends and actual profits is the case of the Union Minière in the Congo. Investment in 1939 equalled about $800 million; for 1953, the estimate was almost $2 billion.[1] Of this "half was from non-African investment and the other half was ploughed back profits".[2] Later, in a 1958 study, the Banque Centrale calculated that the companies were paying dividends averaging between 11.6% and 13.1% during 1951-56. Adding to this the "allocation to reserves" (undistributed profits), earnings ran at 30-35% during the 1950's.[3] However, this by no means completes the picture of all unrealized hidden profits, which were kept closely guarded commercial secrets. The lucrativeness of this "mineral scandal" undoubtedly played its part in the intense power struggle for control of the country after independence.

The significance of such exorbitant profits lies, to begin with, in the degree of exploitation, which leaves little over for development. Moreover, by concentrating in extractive industries, foreign capital in Africa—with the notable exception of South Africa, a colonial power in its own right—tends to perpetuate a one-sided economy and retard industrial development. By and large, the countries not making satisfactory economic progress are being hindered not least of all by foreign capital from re-structuring their economy to make broader and better use of their own resources. And the more lucrative the profits, the more tenaciously imperialism seeks to hold them in its political and economic grip.

---

[1] *According to the Banque Centrale du Congo Belge Bulletin,* August 1955, cited by A. Kamarck, op. cit., p. 194.

[2] Ibid.

[3] Ibid.

It is not accidental that US private capital is concentrated in by far the biggest industrial and, at the same time, racist country in Africa. (See section: US Partnership in Social Oppression.) With about 6% of the population of the continent, South Africa produces about 1/4 of the continent's gross national product and 2/5 of its industrial output. If its minerals and their profitability were the initial attraction of American monopolies, their economic stake soon became inextricably intertwined with more far-reaching political, military and social considerations.

The US share of foreign assets, which is about 1/5 of Britain's[1] and interwoven, has been growing uninterruptedly in the postwar period. This contrasts with the slowdown in investment from other imperialist powers in the late 50's and early 60's.

US monopoly capital has been encouraged by Washington's aggressive postwar foreign policy and by British and South African monopolies to buy into and underwrite the colonial and racist system in southern Africa, not without the promise of fat profits in the bargain. In 1947, Newmont Mining Co. (Morgan group) and American Metal Climax (Hochschild-Lehman interests), for example, were invited by the South African government to buy at what soon proved to be the low price of $2.5 million (estimated worth in 1964 —$80 million) the Tsumeb Mining Co., the largest enterprise in South West Africa (in effect, a colony of South Africa), and soon to dominate base metal mining in this part of the continent. To participate in this venture, Secretary Byrnes resigned from the State Department to become a director of Newmont, and the Dulles' law firm (Sullivan and Cromwell) has since represented American Metal Climax.[2]

In 1957, an even more important step in US expansion occurred when Charles W. Engelhard of New Jersey acquired

[1] The UN Committee on Apartheid reported of $4.4 billion in foreign assets in the mid-sixties: Britain held 60%, the United States—11%, Switzerland—6%, and France—4%. *Quarterly Economic Review*, London, October 1966.

[2] In 1972, 15,000 Namibian mine workers were striking against foreign and domestic exploitation maintained by South African military and police forces. Workers received approximately $300 a year,

at a low price control of the $300-million Rand Mines group, one of the big seven trusts which run the main industry of South Africa—gold. The financing of Engelhard's enterprises, as well as all South Africa government bond flotations, has been handled by the investment banking firm of Dillon Read and Co. (headed by former Secretary of Treasury Douglas Dillon). The tie-up between Washington's officialdom, US foreign policy and a growing economic stake, particularly in gold in South Africa, is instructive.

During the rising tide of the national-liberation movement, South Africa, following the Sharpeville massacre of March 1960, underwent a severe financial crisis in 1961-62. Both symptomatic of and aggravating the crisis was the fact that foreign capital fled the country and little new funds were available.[1] It was at this point that the United States and US-influenced international financial institutions played a key role. The IMF allowed South Africa to draw 75% of her $150 million credit in the space of a year. In September 1961, Rand Selection Corporation (in which Engelhard has an interest) obtained $30 million from American sources. An Italian consortium made a loan of $9.8 million in October 1961. In December, the World Bank loaned $25 million, the Deutsche Bank—$9.8 million, First National City Bank—$5 million, and a US banking consortium granted a credit of $40 million (arranged through Dillon Read and Co.).

Such direct financial support, paralleling imperialist political and military actions particularly in the Congo, helped to create the general conditions and atmosphere in which the South African economy was stabilized and "confidence" restored.

Thus, in March 1964, the *Business Digest of South Africa* was able to report that the government was not drawing on the revolving credits of $50 million available from American and West German banks.

The following year, foreign investment began to grow again, with a total net inflow of capital from abroad in the 1965 financial year equalling 270 million rands (202

---

plus food and shelter, while AMAX and Newmont made visible profits of $15 million from Tsumeb in 1971.

[1] See *Africa Today*, January 1966.

million rands from private sources),[1] in sharp contrast to the previous six years when there was a net capital outflow. This was even of greater political than economic significance. From the strictly economic standpoint, South African domestic capital formation is sufficient to finance a 5.5% growth rate.[2] South Africa felt, it was reported,[3] that the greater the amount of foreign-held assets, the greater would be the interest of foreign powers in restraining economic sanctions against her. This hope was based in the case of British and American monopolies largely on their big profits from certain minerals.

Through investment, trade or both, US big business has an important stake in South Africa's gold, uranium, vanadium, chromium, asbestos, nickel, copper, diamonds and antimony. The Charles W. Engelhard interests alone control nearly 15% of the country's gold production and 20% of uranium. Through their various directorships, they play a part in mineral policy decisions of the producers of two-thirds of South Africa's gold and uranium. Engelhard fabricates all of the US platinum supply (mostly imported from the International Nickel Corp. in Canada, but 28% from Britain to which much of the South African output flows). The late Mr. Engelhard, the prototype of "Mr. Goldfinger" in the literary works of Ian Fleming, was a personal friend of President Johnson and acted at times as a US diplomatic representative to southern Africa.

US monopolies' interest in South Africa's metallurgy was extended to ferrous metals when the Anglo-American Corporation initiated a $140 million high-grade steel project in 1965, with much of the output to go to the United States. With the labor costs in steel running to about 60%, South Africa produces the cheapest steel in the world—considerably below US prices. Thus, we find, Eastern Stainless Steel Co. of Baltimore together with Rand Mines formed the Southern Cross Stainless Steel Co. in 1965 to manufacture in South Africa.

The predominant flow of US capital to South Africa is predicated on the overall level of profits which is steadily

---

[1] *Quarterly Economic Review,* Economic Intelligence Unit, London, October 1966. (1 rand=$1.40).

[2] *Africa Today,* loc. cit., p. 9.

[3] *Quarterly Economic Review,* loc. cit.

high, as it is based on the double exploitation of an industrially developed capitalist country operating on the economics of apartheid. This is especially applicable and relevant to mining, and most broadly to gold, which alone employed about 380,000 Africans, segregated, mainly migrant, contract laborers, and 45,000 Europeans in 1963. As pointed out by the South African economist D. H. Houghton,[1] there was no rise in the extremely low wages of African mineworkers in the 25 years between 1935-60. Moreover, the tremendous disparity between non-White and White mineworkers, rather than diminishing, further increased in this period from 1/11 to 1/16, accentuating the degree of exploitation of the main labor base.

On this basis, it is not surprising that industries generally show profits ranging from 25% upward. In 1962, for instance, the reported average net profits to net worth ratio for US firms was 25%, rising in 1964 to 27%.[2] Annual rates of profit officially reported in manufacturing[3] in South Africa were: 19.7% in 1961, 24.6%—1962, and 26%—1963. Even higher profits came from the country's gold mines, which recorded in 1963, for example, working profits of $378 million[4] out of a $960 million output—or 40%, one-third of which the Government was able to siphon off in taxes. Such high profits representing the visible portion of corporation profits—distributed and undistributed, are typical and obtainable from annual and other published reports.

Since undistributed profits are generally ploughed back, they together with various forms of hidden profits are partly reflected in the growth of the market value of stock. Thus, in the 50 years since the founding of the largest of the 7 big

---

[1] *The South African Economy,* London 1964, pp. 161 and following; *Hearings, Subcommittee on Africa, House Committee on Foreign Affairs, US Congress,* March 1966, Part I, Testimony of V. McKay, Johns Hopkins University, p. 87.

[2] *Business International,* March 6, 1964, cited in *Africa Today,* January 1966, p. 9. The return on "raw" investment was 13% compared to a world average of 7.7%, *South African Summary,* March 12, 1965, Information Service of South Africa, N.Y.

[3] In the motor industry, Ford and General Motors have subsidiaries in assembly plants and component production, and together accounted for half the vehicle sales in 1963-64. The rubber industry is dominated by General Tire and Rubber, Firestone and Goodyear, along with Dunlop Rubber Co. of Britain.

[4] *Africa Today,* loc. cit., p. 22.

mining-finance complexes in South Africa, Anglo-American Corporation,[1] its initial capital of £1 million had grown to £293 million by 1966, or about 12% per year. Moreover, it was by this time controlling companies with over £600 million assets. Its investments in some 221 other companies, which constitute the major portion of its assets and source of earnings, were estimated to be £210 million at 1965 market prices—a threefold rise since 1956, or about 22% per year.[2] This was confirmed in the 70% rise in the market value of its stock in the three-year period 1966-68, a considerably faster rate than previously.

To distributed profits and increased value of stock, moreover, must be added unknown sums funnelled off by the controlling interests, which raise the profit rate even higher and have contributed additionally to the formation of the fabulous fortunes of their multi-millionaire owners[3].

Noteworthy is that the triangular South African, British and US monopolies' relationship to gold as a commodity and source of profit, which is on the whole a centripetal force, was not paralleled by the British and US governmental policy to South Africa with respect to gold as a reserve backing for sterling and the dollar. The latter relationship became an acute problem for Washington in the early sixties mainly arising from the $35/oz dollar-gold ratio fixed in the mid-thirties, which had become unreal. South Africa and the producers of the precious metal were seeking upward revaluation of gold, or its corollary—dollar devaluation, which the United States opposed.

Although the dollar-gold crisis is a complex problem involving US domestic and global relations which requires a study in itself, it represents such a deep-going antagonism that it cannot be by-passed without at least pointing to some of its elements affecting South Africa. For a decade, the crisis has been aggravated by the US balance-of-payments

---

[1] H. F. Oppenheimer is Chairman of the Board, with the majority of the directors South African, a smaller number—British, and one American (Engelhard). Annual Report in *African World*, July 1966. The company's earnings (as percent of total income) were from gold—41%, diamonds—18%, copper—17%, industry—10%, coal—6%, others—8%, *The Economist*, October 29, 1966.

[2] Ibid.

[3] The family interests of Anglo-American, for example, are held through E. Oppenheimer and Sons (Proprietary) Ltd., a private company which does not publish accounts.

deficit leading to a steady gold outflow, the basic causes for which turn on large capital export (earning high profits abroad), paralleled by inflated military expenditures and aid programs.

With the resultant gold drain leaving ever smaller margins of gold backing for US currency, superficial governmental measures to overcome the deficit and halt the gold outflow proved unsuccessful. Yet Washington was unwilling to trim its foreign economic and political policies, although it was prepared to pay off a relatively small percentage on foreign-owned or internationally-held gold stock to preserve dollar stability. When this alone did not avail, however, the United States took an adventurous step in 1968 to avoid currency devaluation by slipping out from its dependence on gold backing for the dollar and creating a two-tier system.

That this did not succeed was evidenced by the more acute and complicated dollar crisis in August 1971, which Washington sought to overcome in December by eliminating one complex of contradictions—a number of currency disproportions—through dollar devaluation paralleled by insisted-upon rival currency revaluation. Although this served to alleviate temporarily some of the contradictions with US rivals, the token devaluation of the dollar in terms of gold by 7.9 per cent was far from bringing it into a realistic ratio to gold as advocated, for example, by French banking representatives. However, since the continued cheap production of gold is made possible essentially on the basis of the cheap labor under apartheid, the more fundamental causes of dollar overvaluation and instability remain.

If US direct economic interests in South Africa have become sizable as compared with those of Britain, this is not the case in Rhodesia where South African and British interests are overwhelmingly predominant. The latter's assets were estimated at £50 million in 1965.[1] Of the 100 largest British companies, 45 have subsidiaries in Rhodesia and cover the complete range of its industry and trade. The country's main trading partners before usurpation of rule by a minority in November 1965 were Britain, Western Europe and South Africa. Since "independence", with

[1] *The Economist*, October 9, 1965; *Labour Research*, London, January 1967.

Britain and South Africa in the forefront of economic relations with Rhodesia, Washington deliberately has stayed in the background to avoid the political opprobrium of close association with racist rule.

The foreign investment—South African, British and US—which has controlled Zambia's copper industry, second only to the US output in the capitalist world, has been estimated at around £300 million[1] in the early sixties. Copper accounts for some 95% of Zambia's export earnings and about two-thirds of government revenue. During the industry's short life of about four decades (but mostly in the two postwar decades) it had produced by the mid-sixties about 8.5 million tons of copper valued at about £2 billion at 1965 prices. The undistributed profits from such sales have gone to make up the great part of the present market value of the investment. Despite this ploughing back of profits, the bulk has been distributed as dividends,[2] or in other forms. Distributed profits in the 10-year period 1954-64 totalled £259 million. Moreover, during the years up to 1963, the company which had managed to gain paper "rights" over Rhodesia's minerals, the British South Africa Co., had received royalties amounting to £160 million gross, or £82 million net.[3]

Ownership of the mines has been in the hands of two interlocking groups. The larger, Anglo-American Group is controlled by the Anglo-American Corporation of South Africa, Ltd., which comprises mainly South African financial interests and also British (the US financial interests are from the Newmont Mining Corporation, with which Oppenheimer has for years been associated). The other major group, Roan Selection Trust Ltd., has as its largest shareholder American Metal Climax Inc., with 46.1%;[4] an additional 40%[5] is in the hands of other US companies. The $45 million (book value) AMAX investment tells little of the present size of its holdings and profits. More indicative of present magnitudes is that the two big groups together have

---

[1] And about £500 million more recently (*Morning Star*, August 22, 1969). See Richard Hall, *Zambia*, London, 1965, p. 265.

[2] President Kaunda, in announcing the government's taking control of the industry on August 11, 1969, declared that the companies have been distributing 80% of their profits.

[3] *Zambia*, op. cit., p. 230.

[4] Ibid., p. 264.

[5] *Business Week*, November 12, 1966.

a capital equal to the entire revenue of Britain in 1910.[1] US monopolies have maintained their strong position directly through Roan Selection Trust and indirectly through interlocking directorates with Anglo-American and the British South Africa Co. (now merged as part of Charter Consolidated).

It is not surprising that the United States has a big interest and stake in Zambia's copper. Since the early 1920's, new prospectors and financiers came from across the Atlantic because the United States had in the first two decades of the century asserted financial and technical supremacy in the world of copper. Chester Beatty, who became a British citizen in 1913, remained closely connected with leading American mining corporations (as well as with Oppenheimer), and helped form Rhodesia Selection Trust (renamed in 1965, Roan Selection Trust). Although the United States is by far the world's largest copper producer, US demand has long outstripped local supplies, and it has imported to meet roughly one-third of its domestic needs, chiefly from American monopoly-owned mines in Chile. Britain, on the other hand, is entirely dependent on imported copper, having ceased to be a producer 50 years ago. For US monopolies, therefore, Zambia's copper has been important not so much as a supplementary source, but to prevent the rise of a world competitor, which could "disturb" prices and cut into profits, and its cheap labor is also useful as a counterweight to American workers in the copper mines of the southwest.

In the Congo, foreign capital was directed largely toward the country's big prize—the mammoth and lucrative former Union Minière du Haut Katanga, which controlled an estimated $4 billion in assets (about $800 million of which in the Congo) in the 60's. The struggle was highlighted at the time of the achievement of independence in 1960. Up to then, the main shareholders of UM were: the Belgian government and private interests in Le Comité Spécial du Katanga, with a controlling packet of 25.1%, 2/3 of which were held by the colony (i.e., Belgian government) and 1/3 by the Compagnie du Katanga (in the hands of Société Générale de Belgique, which also held an additional 4.5%

---

[1] See B. W. Smith, *The World's Great Copper Mines*, London, 1967, p. 16.

of UM shares); and the South African and British interest in Tanganyika Concessions,[1] with 14.5%. As compared to the Belgian, or even South African or British investment, the US stake[2] in the Congo was minor—about 8% of the shares of Tanganyika, or 1.3% of UM.

With the dissolution of Comité Spécial on June 24, 1960, it did not turn over to the Democratic Republic of the Congo the agreed upon 2/3 of its packet, or 16.7% share of UM. Instead these shares were handed over to Companie Financière du Katanga in the secessionist state, i.e., kept in Belgian hands.

If the struggle for possession of this packet and control had been largely in the economic sphere up to independence and in the early 60's, it became largely a political battle within the context of control of the Congo by the middle of the decade. Other factors—political, financial and military—had entered the picture: the financial costs of quelling the patriotic forces, an estimated 60% of which were borne—not altruistically—by the United States; US military and budgetary support in the post-UN phase, and concomitantly Washington's strong influence in the central government.

The US-Belgian relationship is perhaps the focus of the struggle for control, politically—especially with respect to the Kinshasa government, and economically—essentially one of rivalry in the Congo, as opposed to a growing centripetal force within Belgium itself. By 1966, for example, the United States was by far the largest foreign investor in Belgium,[3] and by the close of 1969 the investment totalled $1.2 billion—proportionately as high as the record level of US capital in Britain.

The culmination of this struggle was the seizure by the Kinshasa government of UM assets on January 1, 1967.

---

[1] Tanganyika Concessions (headquarters shifted to Salisbury in 1961) is also partners with DeBeers in Anglo-American Corporation, and owns Benguela railroad across Angola (used since 1960).

[2] Thus, one evaluation of US interests is as low as $20 million, or about 1% of an estimated Belgian $2 billion. (*Le Monde*, January 6, 1967). An official US estimate was $25-$30 million in the same period.

[3] In 1965, foreign investment had reached a record $360 million, or double the 1964 figure, of which 90% was US. (Monsanto Chemicals, Atlantic Polymers, Caterpillar, General Motors, Esso Research.) *The New York Times*, June 14, 1966. Furthermore, US banks, e.g., New York Chase Manhattan Bank, are growingly linked with Brussels.

Subsequent behind-the-scenes hard bargaining in negotiations for a deal has not interfered with continued production and profits.

WEST AFRICA

In West Africa, significant amounts of US private capital have been exported to Liberia, Nigeria, Gabon, Ghana and Guinea.

Liberia, in particular, long a major African attraction of US investment, shows how the US combined economic ties have served to exercise a predominant hold on a country. With 90% of the country's population living on the land, and much of the farming on a subsistence level, US capital in the plantations, chiefly rubber, has provided the main sphere of wage labor, source of foreign profits, and important export product to the United States until overtaken by iron ore in the early sixties. Beginning in 1926, Firestone was operating 100,000 acres by 1967, with the highest annual rubber yield in the world; and Goodrich was second with the 50,000 acres planted in 1955 and producing since 1963. In contrast, the local bourgeoisie had about $2 million invested in 1960, which was less than 1% of the total foreign investment.

Although iron ore deposits have been processed in Africa for centuries, major exploitation did not take place until the 1960's. The United States, having used up its highest grade ores, imports about one-quarter of its consumption from Canada, Latin America and Africa. Liberia, the main African source, is the largest producer on the continent and third largest exporter in the world. An early postwar corporation was the Liberian Mining Co., financed by the Republic Steel Corporation, in which a leading role was played by Edward Stettinius, former Secretary of State.

Growing needs for iron ore resulted in the formation of the largest single mining monopoly in Africa, the Liberian American-Swedish Minerals Co. (LAMCO), a $300 million joint investment, which produced about 10 million tons of Liberia's 17 million ton output in 1966. With 50% of the company owned by the Liberian government and 50% divided between Sweden and the United States, the latter's share was estimated at over $205 million in the mid-sixties.

By influencing the establishment of, and gearing agreements to a low world price on the country's high grade ores, foreign capital is able to hide its high profits and leave little to the country for its own development. Thus, instead of taxing gross profits in the ordinary way and levying a flat royalty per ton reflecting rate of production rather than world prices, the Liberian government is allocated a share of declining visible profits:[1] 37.5% as dividends on its own stock and 12.5% as a tax on the share of profits going to Bethlehem Steel. But prior to distributing profits, LAMCO is writing off its capital at $15 million per year for loans and interest to banks, etc. In addition, substantial sums are set aside as "special reserves" (undistributed profits) and "equipment replacement" (depreciation). The first category and part of the second contribute greatly to increasing the market value of stocks, which are of benefit mainly to the Swedish and American stockholders. In addition, the low prices on iron ore are of benefit to the purchaser of iron ore—Bethlehem Steel Corporation.

The country continues to be structured as a primary producer and dependent economy, with trade overwhelmingly oriented to the United States. Roads and railroads to carry iron ore and other raw materials to the coast are financed by loans from US or international credit institutions. Their effect is to facilitate the draining of the country's resources, with little contribution to its economic growth. The same is true of the ports, including the deep water berths at Monrovia, which is also designed as a naval station for US vessels. Flying flags of convenience because of low registry rates, foreign vessels—many of them US—are listed as belonging to the Liberian merchant fleet, making it on paper the largest in the world with a total of 22 million gross tons in 1967.

Nigeria, a key populous state, became soon after independence a target of Washington-encouraged US private investment. By 1967, it was estimated at $200 million. Although the flow initially was into commerce, manufacturing and banking, the attractiveness of oil has since then outweighed other branches. By 1967, Nigeria was in third place in Africa (after Libya and Algeria) and among 11 principal world exporters. Britain had by far the largest investment, with Shell-British Petroleum[2] holding an estimated 85%—

---

[1] See *West Africa*, February 3, 1968.
[2] See the British Labour weekly *Tribune*, August 2, 1968.

about £200 million; Gulf Oil Corporation had about 10%—£25 million; and French interests—about 5%. Earnings of the oil companies were almost £100 million in 1966.

With the rapid rise in oil output, concentrated in the eastern region, the distribution of revenues between the Federal and regional governments became a big issue in the 1965 elections. Coming on top of tribal hostilities, it may well have played a background role in the coup of Ibos in January 1966, which was followed by a counter-coup and bloodshed. Although it is problematical whether the breakaway of Biafra in June 1967 was fostered by the oil monopolies, they undoubtedly complicated and drew out the war. Thus, initially Britain sought to keep a foot in both camps, but then she cast the die and decided to make revenue payments to Lagos. Indications are that French and Italian firms, in the hopes of getting larger concessions, advanced large credits to Biafra, and that France also had a hand in the Ivory Coast recognition of the breakaway government. The United States, perhaps, played the most ambivalent role, between an official policy for the Federal government and indications of unofficial support for Biafra.

In Ghana, prior to the February 1966 reactionary coup, the non-capitalist path of development incorporated a large measure both of planned domestic investment and of invited foreign capital (in agriculture, power and irrigation, fertilizers and industry). By 1967, the US investment was estimated at $170 million—quite substantial for such a small country.

The seven-year plan launched in 1964 aimed to eliminate unemployment, to alter the colonial structure of production and to mesh with the Pan-African economic community. In an effort to continue to utilize government and private sources, the plan foresaw an investment of £G1,000 million: government—£G476 million, and private—£G540 million.[1] About one-half of the government investment (£G240 million) was to come from foreign loans and grants. This, together with £G100 million of new foreign capital, including £G60 million from the American Volta Aluminum Corporation (VALCO), made one-third of the total investment to consist of foreign capital. It was a bold attempt to

---

[1] W. Birmingham, I. Neustadt and E. N. Omaboe, *A Study of Contemporary Ghana*, Vol. I, The Economy, London, 1966, pp. 453-57.

grapple with an open question—whether foreign capital would be satisfied to cooperate, and to restrict itself to purely economic activities, in a state pursuing progressive social aims.

Since the United States imports about five-sixths of its required aluminum-bearing ores, it is not surprising that the major US investment was in the Volta project (dam, power station and aluminum smelter), which had been considered a "calculated political risk" by President Kennedy. With the aid of US official capital, Washington had encouraged Kaiser in this estimated $300-million joint venture. In addition to the relatively small VALCO investment of $32 million, external financing was to be chiefly by the US Export-Import Bank ($96 million) and World Bank ($47 million). VALCO was to get cheap power and aluminum. The US Development Loan Fund guaranteed the private investment. And even if the smelter were nationalized, the United States, according to agreement, could market through VALCO its claim on the Ghanaian government for aluminum. Hence there was little financial risk for US interests. On the other hand, even bigger economic gains were expected by Accra. By singular coincidence, however, shortly before the dam was to be opened, the Nkrumah government was overthrown.

US capital also was reluctant to enter Guinea without concomitant Washington support in the immediate post-independence years. In early 1964 the situation altered following a $35 million loan.[1] Moreover, by the close of the 60's US government specific risk guarantees had been concentrated to a remarkable extent in Guinea, comprising over one-half of the total guarantees for the entire continent. This political-economic umbrella over private capital could well be described as a continuation of the Kennedy policy of playing for the long term, "to stay in close, keep working and wait for the breaks".[2] This policy in black Africa in the early 60's, according to Schlesinger, had "its most notable success in Guinea".[3]

---

[1] See В. В. Рымалов. *Распад колониальной системы и мировое капиталистическое хозяйство,* М., 1966, стр. 356-57 (V. V. Rymalov, *Disintegration of the Colonial System and the World Capitalist Economy.*)

[2] Schlesinger, op. cit., p. 565.

[3] Ibid., p. 567.

The main foreign role in mining operations was assumed by the United States. The International Consortium FRIA was organized under American influence—Olin Mathiesson (US) and Pechiney (France)—and Harvey Aluminum together with the Guinean government began to work the previously nationalized bauxite resources in Boké. Since then international loan capital also has entered in support of new major bauxite mining operations.[1] Revenue from FRIA has made up more than one-half of the Guinean government's income, and represented 65% of the earnings of the mixed company. Furthermore, US firms contracted to construct a factory turning out aluminum products, under control of the Guinean government. The latter, in turn, undertook not to carry out nationalization for 75 years.

US companies are also active in vital transportation, e.g., Pan American Corporation and Mack Truck (with a monopoly in the sale of tractors).

In contrast to Ghana and Guinea, however, US capital has been less able to penetrate West African countries with strong political ties to France. In mineral-rich Gabon, in which French capital predominates, US private capital has joined the former in a secondary relationship in a number of mining companies. Thus, in Comilog (Companie minière de l'Ogué), French companies hold 51% and the rest is in the hands of US Steel Corporation (Morgan). In Somifer (Société de mines de fer Mékambo), Bethlehem Steel Co. holds 50%.

NORTH AFRICA

US investments in North Africa, mainly in oil, are closely linked with the Middle East and have vast global ramifications, e.g., from exploitation to transportation and refining. Thus, with an estimated investment of at least $4.5 billion[2] in production facilities in the Arab nations in 1967, US monopolies had an additional investment of more than $18 billion in the so-called "downstream" facilities—tankers, terminals, pipelines, refineries, largely in Europe's

---

[1] In September 1968, for example, the World Bank made a loan of $64.5 million to Guinea, supporting agreements made with 7 aluminum companies.

[2] *The Wall Street Journal,* June 12, 1967.

petroleum and petrochemical industry, and market outlets. However, only a few highlights of the basic investment will be treated here.

North Africa and the Middle East, perhaps the most important regional oil complex, reached 40% of the world output of 2,130 million tons by 1969.[1] This, moreover, represented not only the fastest rate of increase of any region during the 1960's,[2] but also the greatest potential for the 70's. North Africa's growth in output, unprecedented in the history of petroleum, was attributable mainly to Libya, which became the region's fifth largest producer in the first half of the 1960's, and then overtook Iraq and Kuwait to become the third largest after Saudi Arabia and Iran by 1968. (On a world scale, the United States was first, but with a slow rate of expansion and a big indicated depletion of reserve during the decade; while the USSR in second place had a very rapid growth rate and indications of large untapped reserves.)

Having achieved a dominant position in Mideast foreign investment in the decade following the war—the United States wholly owns the Arabian-American Oil Co. (ARAMCO) in Saudi Arabia, whereas Britain's stake is primarily (the United States—secondarily) in Iraq and the Persian Gulf—US monopolies extended westward in North Africa after the Suez crisis of 1956. In the decade to 1967, US monopolies[3] gained overwhelming control (about 9/10) of Libya's cheaply produced, high-quality, extremely profitable oil, increasing their interests about 18-fold to constitute one-fifth of the total US investment in Africa, and rising to one-fourth by 1970. (See section "Investment", subsection "General".) US investments in Algeria,[4] mainly in oil (total

[1] *Petroleum Press Service*, London, January 1970, pp. 5-6.

[2] In the 60's, the Middle East and North Africa increased output 3.4 times, the Socialist countries 2.5 times, and the Western Hemisphere showed only a slow expansion—the United States only by restricting imports and Venezuela dropping from first to fifth place. Ibid., pp. 5 and 40.

[3] The companies producing the bulk of the country's output were Esso, Oasis, Mobil, Amoseas and Occidental.

[4] US companies were not interested in being subordinate to the State's controlling interest. (*Business Week*, April 12, 1969.) The Soviet Union, on the other hand, was helping to build the $400 million Annaba Steel complex to produce 400,000 tons of steel in the early 1970's. The French were building a $50-million fertilizer plant and a $190-million liquefied natural gas plant.

output was about 1/4 of Libya's), increased but were small—largely in French-dominated oil concessions and in marketing outlets. Although the UAR's oil production was small (about 1/10 of Libya's) before the June 1967 war, with the Sinai peninsula fields operated either by the government or jointly with ENI of Italy, US firms had been granted concessions in the Gulf of Suez and the Western Desert (Pan American Oil Co., a subsidiary of Standard Oil of Indiana), in Western Egypt (Phillips Petroleum Co.) and were negotiating for a concession in the Sinai. The country's land and offshore potential reserves, which were still undetermined, were arousing the interest of the United States and Israel.

In the search for additional sources of profitable crude oil for the vast petroleum and petrochemical industries of Europe and the increasing needs of the United States, American and European monopolies' interest in North Africa (as well as West Africa) and its offshore reserves was heightened by their location west of the Suez Canal especially in countries amenable to foreign political influence. This could be used as a counterweight to, and in bargaining with, the Middle Eastern countries. On the eve of the June 1967 war, the five countries of North Africa and offshore showed an estimated 25% of world reserves, second to the 40% of the Arab Middle East. As a result of the political changes in the wake of the Israeli-Arab war, however, the feasibility of North Africa being used as a counterweight dwindled and then changed into its opposite: Arab oil had to be considered as a whole comprising two-thirds of world reserves,[1] counterposed to foreign monopoly exploitation.

The region's importance to Washington was further highlighted by the fact that the domestic crude oil reserves of the United States were not keeping up with its demand despite a protected market and high domestic prices. This included the prospects of Alaska's output, which was estimated at about 2 million barrels per day by the mid-1970's. By then, US demand was estimated to rise by 3-4 million barrels per day from its 1970 level of 14 million barrels per day, showing a need for increasing imports from abroad.

---

[1] *Le monde diplomatique*, juillet, 1969.

To conclude, US capital was exported more rapidly to Africa in the 60's than in the world as a whole largely to supply the mushrooming appetites of its monopolies at home and their affiliates in Europe—for petroleum and minerals. Moreover, US African investment, which continued to increase in the early 70's at more than 10% a year, was extracting profits significantly higher than world levels, and if American monopolists and financiers were somewhat disconcerted by the wave of nationalization in a number of countries, it was not restraining investment. Perhaps, this was also attributable to the fact that the total losses in the developing countries of the investment insurers of such countries as the United States, the Federal Republic of Germany and Japan were merely one-tenth of 1% of the value of the insurance cover contracted for—and these, moreover, were indemnified.

As for the results, after two decades of increasing US (and other imperialist) private investment in Africa, no significant decrease could be recorded of the relative size of population living in the subsistence sector,[1] or on the other hand, increase in the manufacturing sector. Africa, therefore, had good reason to be looking for real aid to assist it in its unequal battle for economic development.

AID—POLITICAL-ECONOMIC COMPOSITE

BILATERAL

Whereas Africa looks to aid mainly for economic development, US foreign aid, on the whole, reflects in microcosm more graphically perhaps than any other single sphere the political-economic composite of US foreign policy. The major aim of both military and economic funds is officially to "promote U.S. national security"—in effect, to wage the cold war against the Socialist states and to oppose national anti-imperialist movements.

Within this general political framework, economic funds are also intended to promote US private investment abroad and foreign trade, as well as to a lesser degree—economic development. These categories, needless to say, overlap and are sometimes negated: e.g., transport and communications

[1] In 1971, as in 1950, 59% of population (75% south of Sahara).

do have economic development potential and use but frequently are either for foreign military purposes or exploitation of natural resources and profit; countries receive economic aid, which could be useful, but as compensation for bases, which form part of an imperialist network used to brake political and socio-economic progress. Thus, the underlying aim or function of aid is more meaningful than its official nomenclature and therefore the quantitative figures given are, more often than not, subject to serious qualitative qualification.

The political-military emphasis has been clearly preponderant—although with some variations. Thus, US direct military aid, which amounted to about two-thirds of total US aid to all countries in the 1950's, dropped to about 40% in the early 1960's. But, together with related quasi-military or political categories (supporting assistance and contingency funds) still equalled two-thirds of the total. This proportion, nevertheless, was less than in the previous decade (when it was about three-quarters) and reflected the increasing amounts of funds channeled in economic form—a tendency which generally was characteristic of the sixties.

US aid to Africa, although no less political than to other continents, has had a very small direct military component, i.e., has been mostly in economic form. As for content and direction, up to the middle 1950's US funds went mainly to countries in which the United States had established its bases: Morocco (US naval and air facilities), Libya (Wheelus Air Field), Liberia[1] (US base at Bakers Field and naval port at Monrovia), and Ethiopia[2] (US military communications base). Smaller amounts went to finance the extraction of raw materials in British and French colonies. From 1955-1958, so-called economic aid rose from $37 to $100 million per year, with "access to bases" as the official justi-

[1] A graphic example of how little US funds have aided socio-economic development: By 1960, a former US official reported, "in not a single public school was there a library, adequate textbooks, or sufficient instruction supplied". The Liberian government explained that it had decided—to the tune of an $80 million public debt—that its first priorities lay in developing roads, erecting public buildings and laying the groundwork for expanded private investments in rubber and iron. See J. D. Montgomery, *Aid to Africa,* N.Y., 1961, p. 22.

[2] By 1960 military aid was $42.5 million; and economic funds $72.5 of which $6.4 million was surplus food and $27.4 million Export-Import loans. Ibid.

fication for most of these funds.[1] In contrast, only $20 million was expended in 1958 for the improvement of skills and technology.

Increasing amounts of US aid, although without any fundamental alteration in composition, reflected concessions to the rising tide of the African liberation movement. By 1960, US aid rose to $287 million, about one-half[2] of the economic funds consisting of surplus agricultural commodities and Export-Import commercial loans. Immediately following the "Year of Africa", total US aid dramatically increased to over $400 million in 1961, reached a peak of about $550 million in 1962, and thereafter dropped to an annual average of $480 million in 1963-66. This was followed by a decline, roughly paralleling the curtailment of aid to the UAR and an ebb in the African national-liberation movement beginning with the defeat of the Congolese patriotic forces.

In contrast to the bilateral aid of Britain and France, which was targeted rather closely to their former colonies,[3] US aid was not as linearly linked with previous American political-economic ties. The United States spread out in the sixties from those countries with which it had "special relationships" to a number of newly independent states. Consequently, the pattern of US aid is more understandable as a reflection of Washington's continental or regional political-economic strategy (taken in conjunction with the pattern of the international financial institutions).

Of total US aid to all countries averaging $4.6 billion a year in the 1960's (Table III), Africa received, on the whole, less than one-tenth. Thus on a global scale, the

---

[1] *The Department of State Bulletin,* December 28, 1964.

[2] Ibid., January 25, 1965, p. 105.

[3] About 85% of British bilateral aid has gone to Commonwealth areas, and about half of this to dependencies. Only 10% has gone to non-Commonwealth countries and much of this to Jordan and the Sudan, former British semi-colonies. R. F. Mikesell, *The Economics of Foreign Aid,* London, 1968, p. 14.

About 94% of the official aid of France in 1963 went to its overseas departments and territories, Algeria, Morocco, Tunisia and the African Malagasy states. The FRG aid in Africa, oriented to raw materials and trade, was scattered in some 36 countries, with a tendency to be somewhat larger in former German colonies (Cameroon, Tanzania, Togo) and also closer correlated with funds coming from the United States and the international financial institutions.

continent was of less importance in Washington's aid priorities than Latin America or Asia. However, Africa did receive several times more than the absolute amount of US monopolies' investments and trade with the individual countries would appear to "justify"—if taken separately, rather than as part of a complex.

By far the highest regional level on the continent was in North Africa, which received two-thirds of the total through 1964 (Egypt, Morocco, Tunisia, Libya, Algeria). Then followed sharp reductions in US food shipments to Egypt[1] and Algeria after US pressure failed to turn them from their independent political course. US aid to Libya was discontinued, to all intents and purposes, in the mid-sixties when that country's big oil output and revenues began to dwarf the amounts given her in compensation for a US military base and presence. Thus, in 1967, North Africa—in effect, only Morocco and Tunisia—was receiving about 30% of the total US bilateral aid to the continent, as compared with 90% in 1960.[2]

This smaller proportion resulted in part from the above reductions, but also from a steady or increased flow of US funds to certain countries in other areas. In central Africa, for example, the key state Zaïre had come to be one of the biggest recipients of aid, despite—or perhaps because of—its vast and highly profitable mineral wealth simultaneously being pumped out of the country.[3] US aid was paralleled by Washington's influence in Kinshasa. In West Africa, Liberia continued to be the main aid receiver. Ghana, which had received a moderate amount of aid (together with multilat-

---

[1] President Nasser has revealed, for example, how Washington withheld $60 million of grain when his country refused to permit US inspection of Egyptian industry, reactors, etc. The US Assistant Secretary of State then threatened that the United States would supply Israel with still more arms if Egypt turned its propaganda against Washington. *The New York Times*, February 13, 1970.

[2] The purpose of such "support assistance" to Morocco and Tunisia, moreover, remained the same in the mid-sixties, according to the US Assistant Administrator for African AID, about $100 million as "sort of payment for bases". Foreign Assistance Act of 1966, *House Committee Hearings*, March 1966, Part I, Washington, D.C., p. 150.

[3] Thus, for example, production rose to $400 million in 1966 (10% over 1965) mostly in minerals. Agricultural products accounted for only 23% of exports in 1966 (cf. 45% before independence). This gap, resulting from the failure to win over the suppressed national-liberation forces in the countryside politically, was partly filled by US aid.

Table III

US Aid, bilateral ($ million)

	Postwar through 1970	1957	1960	1964	1967	1970
Total all countries	123,134	5,070	4,437	4,811	4,947	3,593
Total Africa	4,939	51	287	474	342	270
	4%	1%	6%	10%	7%	8%
of which:						
UAR[1]	1,127	7	108	194	5	—
Morocco	690	18	61	39	34	64
Tunisia	623	6	55	44	49	49
Congo (K)[2]	364	—	11	40	35	11
Nigeria	251	—	3	25	35	36
Liberia	240	5	8	12	37	—1
Ghana	223	—	2	8	35	2
Libya	206	17	34	6	—6	—
Ethiopia	185	7	7	8	11	9
Algeria	178	1	1	39	11	1
Sudan	94	—	17	10	—2	—2
Guinea	93	—	—	11	7	4
Somalia	72	—	3	6	5	5

Source: Statistical Abstract of the U.S., 1964-71.
[1] Since 1971, ARE.
[2] Since 1971, Zaïre.

eral aid for the Volta dam) under President Nkrumah, suddenly became one of the largest recipients following the reactionary coup in early 1966. US aid and other officials became prominent in Accra. Nigeria, the important oil-rich and most populous African state, was steadily given funds to become the third largest recipient by 1967. Guinea, which received moderate aid, was the object of more than moderate interest[1] for a small state whose progressive policies were not approved by Washington.

In East Africa, which links up with the Middle East complex, Ethiopia continued to be the main target of US aid. Nevertheless, small and strategically located neighboring Somalia became one of the largest per capita aid recipients

[1] And not only in bauxite. US policy was "to keep a foot in the door" through the presence of aid administrators and renewed aid offers. By combined western diplomacy, according to a former US official, Guinea was discouraged from recognition of the GDR. See J. D. Montgomery, op. cit., p. 36.

towards the end of the decade. The Sudan, which also was marked as one of the 10 countries in which Washington planned to concentrate aid in 1967,[1] broke diplomatic relations with the United States after the Israeli-Arab June war and aid was severed.

The amounts and direction of US aid, which are determined by specific class aims and bargaining position, are reflected in its ideological rationale—even if in distorted propagandistic form. (See section "Ideological Forces".) The string of five different changes of name of US aid administering agencies[2] testify to attempts to overcome the political taint attached to US bilateral aid. Further efforts to this end in the 70's are in the direction of lessening the visibility of US aid administration and greater emphasis on multilateral lending.

In contrast, Socialist aid by its socio-economic origin and nature is essentially directed to supporting the political independence and encouraging the industrialization and economic growth of the developing states. By 1960 the Soviet Union was exporting annually machinery and industrial equipment—particularly needed by the developing countries—worth about 1 billion rubles, or twice the amount of 1955. By 1963, Soviet aid totalled about 3 billion rubles, one-third of which was going to the African and Arab states.[3]

Although this sharp rise has been an outgrowth of increased Socialist economic strength, it has not been without sacrifice to the Soviet economy.[4] Generous credit terms are

---

[1] *The Foreign Assistance Program, Annual Report to Congress,* Fiscal Year 1967, Washington, D.C., 1968, p. 30.

[2] The predecessors of the present Agency for International Development beginning in 1948 were: International Cooperation Administration, Foreign Operations Administration, Mutual Security Agency, and Economic Cooperation Administration.

[3] The biggest recipient of these was Egypt (second to India on a world scale), then Ethiopia, Ghana, Guinea, Mali, the Sudan, Somalia, Tunisia, etc. For a comprehensive discussion on this, see В. В. Рымалов, *СССР и экономически слаборазвитые страны* (V. V. Rymalov, *The USSR and the Economically Underdeveloped Countries),* M., 1963, pp. 56 and following.

[4] This is generally acknowledged. See, for example, F. D. Holzman, *Soviet Trade and Aid Policies* in a Columbia University symposium, in which it is pointed out how real cost of aid in the Soviet Union is greater than in capitalist countries (with idle or surplus capital for export) because of the country's full employment, no surplus of funds for domestic investment and the high rate of return. *Soviet-American Rivalry in the Middle East,* ed. by J. C. Hurewitz, N.Y., 1969.

typical: long-term—for about 12 years, and at low-interest rates—2.5% (cf. 4-7% and more from capitalist states). Payment is made out of receipts after commissioning, frequently with the output of the plants constructed or with the country's traditional exports. This helps to provide them with a stabilized market and prices. To accelerate industrialization, credits are largely for specific projects, equipment and technical assistance is provided, while the assembly of plants is performed by local personnel who also receive on-the-job training. Some of these economic, financial and technical features are in sharp contrast to US and other imperialist aid.

How the political-economic composite of Socialist assistance contrasts with that of the United States is well illustrated by the now classical events surrounding the financing of the Aswan Dam. In 1956, when Secretary of State Dulles instigated the withdrawal of the US-World Bank offer to finance the Aswan Dam, the Suez Canal was nationalized with one of its aims being the raising of larger amounts of domestic funds. The imperialist reply to this step was the blockade of Egypt and British-French-Israeli aggressive war in October-November 1956. In the wake of this, Washington also joined in blocking the foreign assets of Egypt, and in July 1957 the United States refused to sell it wheat or buy its cotton. The Soviet Union did both, and also provided oil and assistance. On January 29, 1958, the agreement was signed to finance and build the Aswan Dam, which was to expand by one-third the 2.4 million hectares of cultivable land, increase the National Income by 45%, generate 2.1 million kilowatts of electric power (5 times the existing hydroelectric power output), more evenly distribute water increasing inland waterway shipping by 20-30%. The dam was to pay for itself in 2 years after completion.[1]

Such examples of direct Socialist aid in construction, and supplying equipment and technical assistance[2] have thrown

[1] V. V. Rymalov, ibid. Moreover, Nile River control already has averted what otherwise would have been a flood disaster in 1964 and serious crop damage to output from low rainfall in 1965. See A. S. Gerakis, "Some Aspects of First Five-Year Plan" in *Financial and Development Quarterly,* No. I, 1969, IMF and IBRD.

[2] In 1960-67, for example, the Socialist countries have trained 150,000 qualified workers and technicians in the developing countries; 12,000 on the Aswan Dam and 40,000 in other African countries. In the USSR, professional training has been given to 30,000 and in Czechoslo-

into glaring relief the minimal efforts being made by imperialism in overcoming economic backwardness.

Although economic development is a minor aim of US funds—in fact, if not in words—US officials are indeed concerned, and not without reason, over the widening gap between the advanced industrial countries and the developing states as a potential threat to continued imperialist interests and influence.[1] Consequently, there has been marked recognition since the early sixties that some economic and social development is a necessity.[2] The minimum goal of a 5% annual growth rate set by the UN in the Development Decade was to be achieved with the help of an annual aid flow of 1% of the GNP of donor countries. Quantitatively, however, this was only about one-half met by the United States.[3]

But, percentage of GNP as an indicator would be of greater significance—even in the purely economic framework—if it were not counterbalanced by a substantial outflow taking place from the country. Thus, terms of aid and external debt are of immediate relevance. Yet the United States instead of lowering, has hardened its terms, the average interest rate for bilateral loans, for example, increasing from 3% in 1964 to 3.6% in 1965. (International Bank for

---

vakia more than 1,000 from the developing countries. In 1967-68, there were more than 500 Soviet teachers in Africa. "Financing Economic Development: International Movement of Long-Term Capital and Official Donations, 1963-67," UN, 1969.

[1] See Robert McNamara, *The Essence of Security, Reflections in Office,* N.Y., 1968.

[2] The vast literature on aid abounds with examples, including formulas for measuring such concepts, e.g., Political vulnerability (P_e), i.e., an "inclination to Communism", set forth by Charles Wolf (*Foreign Aid: Theory and Practice in Southern Asia,* Princeton, 1960). His P_e is a function of 3 complex variables—varying directly with a) economic aspirations, and inversely with b) the current standard of living and c) economic expectations. In general P_e according to that author, goes hand in hand with inability to achieve satisfactory rates of growth and development. See, *The Economics of Foreign Aid,* R. F. Mikesell, London, 1968.

[3] It was, at its high point, some 0.55% of GNP in the early 60's (cf. France—1.32%), [According to John A. Pincus, "The Cost of Foreign Aid" in *Review of Economics and Statistics,* Vol. 45, November 1963, p. 364] depending on how Public Law 480 (food) is calculated—at official or world market prices. See, H. G. Johnson, *Economic Policies Toward Less Developed Countries,* Brookings Institution, Washington, 1967.

Reconstruction and Development raised its interest from 5.5% to 6% in 1966, 6.5 in 1968 and 7% in 1970.) The total outstanding African external debt continued to rise, shouldering the debtor countries with an increasingly heavy burden of interest charges and repayments,[1] which took a greater share of their revenue from exports. Consequently less was left for economic growth and development.

This still does not take into account the quality, or internal effect, of aid in continuing by and large the colonial structure of—rather than restructuring—the economy, e.g., US aid essentially for infrastructure for private investment in extractive industries, budgetary support and food surpluses[2] when not paralleled by economic development projects. Of critical importance has been the socio-economic orientation of both donor and recipient.

Thus, US food could be of assistance to countries in need while engaged in industrial and agricultural development, raising technological level and productivity, or balancing the economy. In the UAR and Algeria where US food was being used in this sense, however, US aid was employed as a Damocles sword and was then cut off for political purposes. Its effect in countries which have not embarked on a strong independent course aimed at restructuring their economies has been to act as a crutch and to maintain their dependence.

In Tunisia, for example, where France traditionally had the strongest foreign influence,[3] US aid has moved in beginning in 1957, but especially since 1964 when French aid was cut off in reply to nationalization measures. US aid, which reached a cumulative total of about $500 million in 1967—the highest per capita amount of US aid to any African state—constituted over half of all the foreign aid received by that country. The FRG and the World Bank were addi-

---

[1] Between 1962-66, whereas payments on external public debt for all developing countries grew at an annual average rate of 10%, considerably faster than increase in exports, in Africa these payments rose about 15% annually. *Annual Report, IBRD*, 1966-67, pp. 30-31.

[2] The food component has been particularly large—over 40% in the middle sixties. See *Foreign Assistance Act of 1966*, op. cit., pp. 138 and 153.

[3] From aid and financial ties, trade (wine and other products) to technical and cultural links, e.g., about 3,000 French teachers and technicians in the country. *New Africa*, September-October 1969, London.

tional important sources of aid. US aid has been mainly (about one-half) in food shipments depending on the country's harvest. A bad harvest means complete dependence in a country where 60% of the population is engaged in agriculture and produces a maximum of 75% of the country's needs. Another factor in continued economic-financial dependence was the fact that over 40% of domestic investment was being met from foreign funds in the early sixties.[1] Service payments on external official debt reached 12% of export earnings in 1968—one of the highest in Africa.

In Morocco, too, the United States moved in steadily in the 60's through the medium of aid, and particularly since 1966 when French aid dropped from its annual level of about twice that of the United States down to $2 million.[2] US aid averaged over $50 million annually in the 60's, in the form of budgetary support and surplus foods (at world prices, 6% interest, payable in 3 years). The composition of imports from the United States has hardly changed since 1958—about one-half for consumer goods, 35% for fuel and raw materials, and only 15% for agriculture and industry.[3] Current expenditure in the early 1960's rose continuously from year to year, particularly in 1963 due to hostilities on the border against Algeria. Personnel on the government payroll (including military) rose 25% between 1961-63 with an "unusual reliance on foreign personnel and services".[4] Defense expenditures rose from DH 216 million ($43 million) in 1960 to DH 333 million ($66 million) in 1965,[5] and together with internal security expenditures, constituted about 30% of the government budget. The mounting debt, interest and other payments (including compensation for nationalized lands), it was estimated, absorbed about one-half of the current US and French aid in 1966. This left little for development and restructuring the economy.

In sum, US aid which was heralded in the early 1960's as geared to making an important contribution to economic development is demonstrably more related to US political-

---

[1] Ghazi Dwaji, *Economic Development in Tunisia,* Praeger, N.Y., 1967.

[2] *The New York Times,* January 27, 1967.

[3] *The Economic Development of Morocco,* published for the IBRD by Johns Hopkins Press, Baltimore, 1966, p. 31.

[4] Ibid., p. 35.

[5] Ibid., p. 318.

military and foreign economic strategy than to achi growth and development.[1] In amount—both absolute a a percentage of GNP—it has significantly declined second half of the decade. Its terms have hardened and growing debt and interest charges have increased the burden and dependence of certain countries, e.g., Tunisia, Morocco, Zaïre, Liberia, Ethiopia, Ghana (since 1966), with little visible results in terms of growth.

A general awareness of this has made US bilateral aid increasingly suspect in the developing states, and has led moderate Administration critics, like Senators Fulbright, Church and Proxmire, to oppose aid programs, for example in November 1971, on the basis of their overemphasis of the political-military aspect, to recommend separating the present 40% in economic form from the 60% military and quasi-military to be left to the Pentagon and CIA, and to oppose their "wastefulness"—with the US failures in Indochina and Greece uppermost in mind. Liberal critics of aid programs who are more sympathetic to the plight of the developing countries, such as Gunnar Myrdal and Teresa Hayter (associated with the international agencies), more openly criticize aid because of its stunting and distorting of development. Such reformers, however, are not usually prepared to indict imperialism for its class aims, which are antithetical to those of the developing states and the root cause of the failure of US bilateral aid. Instead, the combination of moderate critics and liberal reformers have constituted—together, incidentally, with "far-sighted" big business and financial leaders—an important factor in Washington's greater emphasis on the international financial institutions.

US MULTILATERAL AID[2]

Although the United States has provided the bulk of its aid funds to the developing countries through bilateral pro-

[1] Thus, we find bourgeois economists now seriously questioning whether aid programs can achieve any self-generating growth even at minimum level. "It is impossible to discern any economic rationale for distribution of aid," writes one author. Countries tend to distribute on the basis of historical, political and commercial relations. R. F. Mikesell, op. cit., p. 269.

[2] Most of the funds provided by the international financial institutions are not considered aid by a wide spectrum of opinion ranging from *The Economist* to the official view of the United Nations, which

grams, of major (and still growing) importance in its strategy are the international financial institutions—mainly the World Bank. Between 1960 and 1964-65, for example, loans of the World Bank to African states rose more than five-fold—from $40 million to $213 million. This level dropped somewhat in the following 3 years, but rose to $345 million in the fiscal year ending June 30, 1969.[1] Thus, from one-seventh of the amount of US bilateral aid in 1960, the Bank's loans and credits rose to one-half in 1965, and surpassed it by the end of the decade. Moreover, while US bilateral aid was expected to continue its downward trend in the 1970's, the Bank planned to treble its lending to Africa (as compared to a doubling of loans to the world as a whole) in the first 5 years of the decade.

Decreasing US bilateral aid and increasing activity of the international institutions are not spontaneous unrelated tendencies, but reflect Washington's policies and ability to implement them. In the World Bank, which was organized and financed initially by the subscriptions of the capital-exporting countries, the United States, with some 28% of the Bank's government subscriptions and 25% of the vote in 1967,[2] is by far the dominant power. Britain is second with 11% of the subscribed capital and 10% of the vote, followed by the FRG and France, each with about one-half of the latter. Only some 10% of subscriptions are actually paid in, e.g., the United States had paid in $635 million in 1967.

Presiding successively over the Bank since its formation have been the representatives of the amalgam of US big business, finance and government: John J. McCloy—from Assistant Secretary of War to IBRD, then to US High Com-

---

excludes loans on commercial terms. This was also recognized by the late President Kennedy, while still a Senator, when he deplored the granting of "inflexible hard loans through the Export-Import Bank and the World Bank with fixed dollar repayment schedules that retard instead of stimulate economic development". Speech of February 19, 1959, *Congressional Record,* 86th Congress, Senate, p. 2484.

[1] See the Annual Report of the World Bank and the International Development Association of the corresponding years.

[2] The subscription mechanism, which is related to voting power, was alleged to consist of the formula: 4% of the country's national income in 1940 and 6% of its annual foreign trade in 1934-38, with a 20% leeway for negotiation. The US subscription and voting percentages have somewhat diminished since the Bank's establishment in 1946, but as the richest country it continues to have by far the most leverage.

missioner for Germany, to chairman of Chase National Bank and director of big corporations; Eugene R. Black—from vice president of Chase National Bank to IBRD, then to director of big corporations and foundations; George D. Woods—chairman of First Boston Corporation to IBRD, then to corporations; Robert McNamara—Ford Corporation to Secretary of Defense, then to IBRD. US and British nationals make up 50% of the Bank's regular professional staff. It would be difficult to find a body which more typifies the US oligarchy and its world finance relationships.

Like in other corporate forms, the controlling voices, with their relatively small percentages of paid in subscribed capital, extend not only over the smaller countries' subscriptions but over the much larger sums used for lending operations (about 3 times as much)—the bonds and notes sold to banks and private investors[1], and the earnings from the loans made on commercial terms. By 1969, the Bank in its 23 years had loaned about $13 billion[2] mostly to developing countries, over five-sixths of which was long-term loan capital at conventional interest rates and terms (the remainder were IDA "soft" loans). This brought it a regular and dependable profit, which amounted to $170 million in 1969 as compared to a $145 million annual average in the previous 5 years. The profitability of the Bank, however, is a quite secondary aspect of its lending operations.

The political-economic aims of and accrued advantage to international finance and the monopolies—especially of the United States as the major capital exporter—are to be sought mainly in the Bank's stated function of acting "as a safe bridge for the movement of private capital into international investment".[3] How the Bank promotes this movement is rather candidly admitted. It advises governments to change "inequitable and restrictive legislation" to attract private capital and service hard loans, it frowns on government

---

[1] About $4 billion, 40% of which was held in the United States, was outstanding in 1969. See interview with World Bank President McNamara, *The Banker*, London, March 1969.

[2] This was both sizable and increasing with respect to the flow of official bilateral funds. Thus, the Bank's loans and credits of $1.8 billion in FY 1969, almost twice the previous year's level, compared with about $6 billion of world total official bilateral funds to the developing countries (60% of which was US). *The Annual Report of the World Bank and IDA*, 1969.

[3] *The World Bank Group in Asia*—A Summary, September 1963.

ownership on the pretext of "management considerations",[1] and it refuses to lend even at conventional terms for purposes which, in its judgment, could be financed by private capital. To this end, the United States turns to the World Bank to establish basic policy criteria for a country's tax structure, the allocating of budgetary resources and pricing policies.[2]

Africa, for example, according to the Bank's President McNamara, must undertake "tax measures" and "choice of projects that might be politically unpopular", and show a "willingness to accept and implement advice from outside experts".[3]

Such advice has been geared to promoting and underpinning profitable foreign investment, mainly that of Britain and the United States, and is reflected in the structure of, and decisions regarding the granting of Bank loans. The general overemphasis on transport is indicative. By the early 60's, of the loans of $860 million to Africa, some 55% had been allocated for transport. By 1967, this dropped but was still high at 43%,[4] with electric power—28%, and the rest for industry, agriculture and education. Transport, electric power and other public utilities, although not unproductive and even essential for commodity-producing sectors, are not in themselves a valid indicator of development (cf., for example, industrialization, higher productivity, larger skilled and educated working class). Like other aspects of the economy, they must be examined in context—for whom and what purpose do they serve—and country-by-country. In Africa with its thin population spread over large areas, investment per capita in transportation is disproportionately high.[5] It is generally allocated for the building of roads, railroads and ports—and like electric power[6]—for the ex-

[1] *IBRD, International Bank: 1946-1953,* Johns Hopkins, 1954, p. 49.

[2] See Statement of E.C. Hutchinson, Assistant Administrator for African AID in *Foreign Assistance Act of 1966,* op. cit., p. 146.

[3] *The Banker,* op. cit., p. 198.

[4] Figures for the world (including Africa) were: a third for transport and somewhat less for electric power. *World Bank and IDA Annual Report, 1966-67,* p. 66.

[5] This has been acknowledged even by the World Bank chief economist A. Kamarck, "The Development of Economic Infrastructure" in *Economic Transition in Africa,* ed. by M. J. Herskovits and M. Horwitz, London, 1964.

[6] Of 6,525 million kw electric power output in tropical Africa in 1957, e.g., 5,125 million were used essentially for mining in the Congo and Rhodesia. Herskovits, op. cit., p. 271.

traction of minerals to be shipped out and processed in the advanced industrial countries, with little focus on African development.

World Bank loans, paralleling British and US investment, also have helped to continue imperialist political control relationships of the colonial period[1] to the present day. This is especially evident in colonial and racist southern Africa. Thus, by the end of 1962, of the $900 million provided Africa by the Bank, over one-half went to southern and central Africa—South Africa, Rhodesia and the Congo (Leopoldville, now Kinshasa). Although this region is already interlinked through foreign investment and trade, the neocolonial pattern has been neither strictly economic nor haphazard. Thus, for instance, the decision to build the Kariba Dam, financed by the World Bank and a consortium of banks and mining companies, in Southern Rhodesia in 1955 disregarded a decision made in an earlier period—when economic considerations were paramount—to construct a dam on the Kafue River in Zambia (then Northern Rhodesia), despite the fact that Northern Rhodesia had already spent £500,000 on preparatory work, was to use most of the generated power for its copper belt, and possessed vastly greater irrigation potential. The decision to construct Kariba visibly was politically made to favor the dominant position of the white settler regime of Rhodesia[2] and to the detriment of African-ruled Zambia. That this bias has continued to be built into the policy of the World Bank is evidenced by the more recent decision to finance the Cabora Dam in Mozambique in conjunction with South Africa, thereby giving similar economic support to the colonialist and racist regimes.

In the second half of the sixties, Bank loans and credits branched out greatly, altering the overall contours. Thus,

---

[1] At that time a prime function of railroad construction was for strategic or administrative control reasons—the British in the Sudan for reconquest of the country, Germany in Tanganyika, Britain from Mombasa to Lake Victoria, France from Dakar. "The record of the U.S. was even worse vis-à-vis Liberia." See Andrew M. Kamarck, *The Economics of African Development*, Praeger, 1967, p. 11.

[2] This was the general policy of British colonialism during the period of federation of the Rhodesias and Nyasaland (1953-64) and in preparation for independence. As a result, Zambia also is dependent upon Rhodesia and South Africa for outlets for its copper exports and the bulk of its imports. For a fuller discussion see R. Hall, *Zambia*, London, 1965.

by 1968, of a cumulative total of $1,702 million[1] (Bank loans —$1,426 million, IDA credits—$276 million), as much as 70% went to countries not in southern or central Africa, also shifting politically to a number of selected target countries. In Northern Africa, for example, instead of the UAR and Algeria which had received loans in the early sixties, funds went to the Sudan (Roseires Dam, electric power, transmission lines, and railways—with 3 US banks also participating), Mauritania, Morocco and Tunisia. In West Africa, instead of Ghana, which had received credits —together with US official and private capital—up to 1962 (and then not again until after the coup of February 1966), Nigeria was given the highest priority, with loans for the Niger River dam, railway, port (Lagos) and roads. In East Africa, emphasis was mainly on Kenya, either separately or increasingly regionally (with a tendency to continuing its predominance over Tanzania and Uganda). An analysis of the policies and country emphasis of the Bank in the second half of the decade reveals a lending pattern oriented and complementary to Washington and London political-economic strategy rather than to African independent economic development.

The growing preference of the US financial-business-government complex for international, as distinct from bilateral, aid is not without basis. Multilateral aid has been of particular advantage to the United States, the biggest capital exporter. With relatively small amounts of subscribed capital, the United States has been able to coordinate and control a flow of loan capital to promote and protect US private investment. By screening its own role within an international body of essentially world capital exporters (but including developing countries as members), the United States can become deeply involved in the touchiest domestic decisions of developing countries without raising charges of economic imperialism or neocolonialism. US foreign policymakers consider this to be of prime importance in dealing with the non-aligned nations.

Before granting loans, moreover, the Bank examines not only the specific project under consideration, but the entire

[1] This included (in million dollars): South Africa—242, Nigeria—222, the Sudan—151, Kenya—124, Ethiopia—98, the Congo (K)—92, Rhodesia—87, Algeria—81, Mauritania—73, Morocco—71, Tunisia—58, the UAR—57. *Annual Reports of IBRD and IDA.*

economy of the country.[1] Such economic intelligence gathering would scarcely be permitted by an individual imperialist country. The implication of entrusting such confidential information to the financiers of world capitalism leaps to the eye, especially when it is recalled how access by the banks to inside information of corporations has played its part historically in giving finance capital a key lever in the industrial world.

Today, coordination of world loan capital goes well beyond the international financial institutions as such in meshing imperialist world policies, by embracing, for example, the Organization for Economic Cooperation and Development, the European Economic Community, consortia, and the IBRD consultative groups (e.g., East Africa). Representatives of these organizations meet weekly (and sometimes daily) to discuss financial policies. The proportion of official development assistance for which coordination arrangements existed in 1968 amounted to two-fifths of the world bilateral and multilateral total.

Evidence of trends in the 70's points to an increase in the scope of activity of world loan capital and its global strategic approach. The World Bank has been planning to expand its borrowing, for example, by raising funds from Saudi Arabia, Kuwait (at a time when the Arab world is looking to them to finance development) and the FRG, and to increase its lending by seeking out countries where loans can be made rather than waiting for applications as in the past. An example of the latter is the establishment of an investment advisory team in the Office of the President of Zaïre. Such initiatives in African and Middle Eastern countries have political-economic implications far beyond the framework of purely lending operations.

TRADE—SHORT-TERM FLOWS AND LONG-TERM PATTERNS

The flows of US trade with Africa in the 60's and 70's although based upon the classical economic patterns of obtaining cheap and needed raw materials, the broadening of

[1] See President McNamara speech in Bond Club, New York, May 14, 1969.

markets, and the consequent derivation of profits, are much more involved. To concentrate attention on trade alone as an independent economic category, or for that matter, solely in conjunction with investment and aid with which it is indeed closely linked,[1] is necessary but not sufficient to explain certain important trends and developments. Moving parallel to political ties and aims, US trade, albeit with fundamental direct and indirect economic motives, is highly political—tending either to reflect and reinforce existing US political-economic relations or aiming to forge new ones. In a real sense, US trade, both private and governmental, is an arm of American monopolies in the making and conduct of US foreign policy.

How are the political and economic aims of US imperialism reflected in its trade flows and patterns? The directions —whether encouraged or retarded—in the past decade (see Table IV) are in themselves revealing, particularly so since the United States did not have a colonial heritage in Africa comparable to that of Britain, France or Belgium, which have continued to trade mainly with their former colonies.

The Table shows that the United States has had substantial trade flows with target countries of different socio-economic systems in most regions and particularly with several in the traditional British sphere. Let us examine a few of US major trading partners and how they have been affected by US governmental policies—political, commercial and financial.

By far the most important US trading partner in Africa has been and continues to be the politically "reliable" and industrially developed racist state of South Africa. Thus, US trade (like investment) represented in 1968 about 30% of its total with the continent. Although US exports and imports were both of major importance, exports have on the whole predominated,[2] with the exception of the critical period 1961-63, when the United States significantly increased its

[1] High rates of profit from foreign investment constitute a mortgage on other economies which then must increase exports not for development but merely to pay profits and dividends. Furthermore, the debt burden of the developing countries amounting to $60 billion, according to McNamara, President of the World Bank, was growing twice the rate of export earnings. *Le Monde Diplomatique,* June 1972.

[2] South Africa's enormous exports of gold, mainly to London, counter-balance what would appear to be an overall balance of trade deficit.

Table IV

US Trade (in $ million)

	EXPORTS to			IMPORTS from		
	1960	1964	1970	1960	1964	1970
Total, all countries	20,550	26,438	43,226	14,654	18,685	39,963
Africa	766	1,218	1,579	534	917	1,111
of which:						
South Africa	277	393	563	108	249	288
UAR[1]	151	268	81	32	16	23
Congo (K)[2]	26	66	62	68	45	41
Nigeria	26	64	129	40	35	71
Ghana	17	25	59	52	74	91
Libya	42	59	104	—	29	39
Liberia	36	35	46	39	48	51
Ivory Coast	—	19	36	—	64	92
Angola	11	11	38	26	55	68
Ethiopia	7	12	26	27	53	67
Algeria	24	53	62	1	5	10
Federation of Rhodesia and Nyasaland	15	23	—	16	25	—
Zambia	—	—	31	—	—	2
Morocco	34	37	89	10	6	10
Tunisia	21	32	49	—	1	3

Source: Statistical Abstract of the U.S., 1965, 1971.
[1] Since 1971, ARE.
[2] Since 1971, Zaïre.

imports, and thereby helped to ease the serious political and economic difficulties experienced by this internationally censured apartheid state. Since then, a continued high level of US imports (plus greatly increased British and Japanese imports) and growing US exports, particularly of machinery and transport equipment, have helped to promote the country's strived-for economic and military self-sufficiency.

The striking growth of South Africa's trade in the face of a world boycott can scarcely be viewed as spontaneous development. It obviously has been made possible by its principal commercial partners—the four big imperialist powers, which account for three-fifths of its trade.

For US monopolies, second only to those of Britain (like in the sphere of investment), South Africa has become more than simply a profitable trade partner. This touches the ex-

Table V

South Africa's trade* (in mill. rands)

	Imports		Exports	
	1963	1967	1963	1967
Total, of which	*1,213*	*1,921*	*919*	*1,356*
Britain	362	497	279	410
United States	204	333	82	108
Federal Republic of Germany	130	239	52	81
Japan	56	115	70	175

* UN statistics.

tremely politically sensitive field of weapons supply. Thus, although France and the FRG replaced Britain after 1964 as the main military suppliers of the country in defiance of the UN arms ban, the sale of US licenses, for example, to build transport and military equipment, has more than mere economic implications. It is helping to put the technology for a "big stick" in the hands of the white minority against the national-liberation movement within the borders of the country and to the north. At the same time, Washington takes pains to hush up such trade, for it cannot disregard its "democratic image" vis-à-vis independent Africa. On the other hand, the growing US trade with South Africa cannot be concealed and must be rationalized. It is not surprising, therefore, to find the United States buffering—even if rather cautiously—South Africa and Portugal in the halls of the United Nations or defending their presence at conferences such as the second UNCTAD at New Delhi in February-March 1968. At the latter, notwithstanding US tactics,[1] the delegates of independent Africa registered their protest against the participation of the continent's main colonial and racist régime by walking out en masse.

[1] The US delegate, Assistant Secretary of State Eugene Rostow, for example, admonished the participants "to concentrate on the difficult practical problems" and not on "political problems". *Le Monde,* 8 février 1968.

No less politically revealing, at the opposite end of the continent, are the second largest regional US trade flows to North Africa. Encountering here two countries, the UAR and Algeria, pursuing an independent anti-imperialist course, Washington flexibly sought to turn government-owned food surpluses to political advantage. Thus, the US aid-financed sales of grain to the UAR for local currency were used as a lever in an effort to pry political and military concessions. When these were not forthcoming, aid-financed trade dropped precipitously. The Israeli-launched June war, it may be noted, came hard on the heels of deteriorating US-UAR political and commercial relations. An abnormality of the latter right up to the war was its one-sided relationship. Thus, the UAR imported most heavily from the United States (about 20% of the total), followed by the Soviet Union, the FRG, Britain and France. But, in exports (1966) the sequence was the Soviet Union, Czechoslovakia, India, China, Italy, the GDR, and the FRG—with no evidence of the United States. The lack of reciprocal imports from the UAR reflected US lack of enthusiasm for helping that country to find markets for its commodities.

US trade with Algeria has been rather similar to that with the UAR both in political aim and economic structure, consisting largely of aid-financed food exports with no commensurate reciprocal imports, but has been on a more modest scale (less than one-half of US-UAR trade on a per capita basis). US exports, which have been second to those of France (Algeria's main trading partner in both directions), declined after the June war but not as sharply as with the UAR. With an eye on US oil investment, Washington, as reported in Congressional hearings, had greater hope of continuing its policy of "keeping a foot in the door", undertaken under President Kennedy at the beginning of the decade.

In contrast to the UAR and Algeria, US trade with Libya, Morocco and Tunisia has reflected the much stronger imperialist political influence and economic ties derived from different combinations of oil investment, bases, bilateral and multilateral aid. The volume of US trade has been steady but also skewed—second in each case to that of the former colonial power in exports, but lower in imports (especially from Morocco and Tunisia). In the aftermath of the Israeli-Arab war, the United States greatly increased its trade with

Libya (whose US-owned oil output goes mostly to Western Europe) in both directions in an apparent bid to broaden its economic and military wedge in the Arab world. But this effort was cut short with the overthrow of the monarchy two years later, when the revolutionary government took steps to bridle British and US influence forcing them to relinquish their bases in April and June 1970, and imposing curbs on investment. Morocco and Tunisia, on the other hand, were being drawn in the opposite direction, towards closer commercial and financial ties with the United States and the West European powers, e.g., both countries signed agreements with the EEC in March 1969 and granted expanded air and naval base facilities to the United States.

The trade pattern of Zaïre is a vivid example of an overlay of post-independence upon colonial relations, and of particular interest because it is the only major African country thus far in which the United States has gained a superior political-military position over the former colonial power. Although US trade is steady in both directions, its relative rank is high (second to Belgium) only in the country's imports—the effect of Washington's being the predominant creditor, food and military supplier of the Kinshasa government. On the other hand, the continued colonial structure of the country's exports (copper—about one-half of the total, cobalt, palm oil, coffee, diamonds, tin, zinc, rubber), which go mainly to Belgium, followed by Italy, France, Britain, the Federal Republic of Germany and then the United States, mirror Zaïre's continuing investment and appendage relationship to Western Europe.

Finally, in West Africa—particularly Nigeria and Ghana—US trade has broken perhaps more new ground than in any other region. The strong impress of US political and economic relations and cross-currents is traceable (see Table IV). Thus, in Nigeria, the early designated major target of the Kennedy Administration in government aid and private investment, US trade evolved in both directions, but much more markedly in exports (second only to Britain), until 1967 when the country became the US third largest African trading partner. This was interrupted by the secession of the oil-rich eastern region, which whetted the appetites and drew the support of US monopolies. The following two years saw a sharp drop in US trade with the federal government until the imperialist- and colonial-backed gamble on separation ended in failure. An expected US trade upswing with the re-united country in the 70's, although encountering a more wary Nigeria than in the previous decade, soon exceeded the prewar levels.

Similarly, Ghana, one of the richest and influential countries in black Africa, was marked soon after independence in 1957 as an object of special Washington interest. The quickly emerging US strong position behind Britain in total trade flowed primarily from the complex of US government-encouraged (and international) credits combined with US-guaranteed private investment in minerals and hydro-power, which had been undertaken by Washington in 1961, as well as from Ghana's biggest and most vulnerable source of foreign exchange—cocoa. The international "squeeze" on cocoa prices in the mid-decade, and the consequent reduced export earnings (despite Soviet purchases), had no small effect in accentuating the country's economic development difficulties and nourishing the soil for the February 1966 coup against Nkrumah. Immediately thereafter, the feverish granting of international credits (previously withheld) helped to boost trade with the United States, with regained profitability as a result of a "miraculous" rise in cocoa prices.

Contrasted with new penetrations in the opened-up British sphere, US trade has found easy sailing in its traditional area of "special interest". Liberia is perhaps the classic example in Africa of US anticipation of the "new colonial era" via the entire economic gamut of investment (plantations, iron, railroad), aid, banks, insurance, shipping, and commerce. In all these interconnected spheres, which are linked up in a certain sense by trade, the United States is the paramount power. In colonial fashion, it is first—followed by the FRG—both in the country's imports (about 40% of the total) and exports (about 30%), taking about one-fourth of its iron ore and nine-tenths of its rubber. The country's traffic is in the world's largest fleet, which is mostly US-owned, and insurance is controlled by two American companies. Within Liberia, the biggest commercial firm is the "U.S. Trading Co." (cigarettes, automobiles, etc.) and "Texaco" distributes its petroleum products throughout the country. The economic and commercial predominance of US monopolies gives Washington an unrivalled political grip, which leaves the country little more than its nominal independence.

Although trade expansion with the tighter post-independence franc zone has proved less spectacular for US monopolies than in Liberia or the British sphere, nevertheless, it is not to be discounted. In a few countries, e.g., the Ivory Coast, the increasing imports of such a new and big customer as the United States are playing a growing role in an expanding but fragile economy, dependent on the export of coffee, cocoa and bananas. Although the franc zone still accounted for over 45% of the country's exports in the middle of the sixties, the dollar zone was second with up to one-third of this. Moreover, US coffee imports approached the level of France's in 1964, although dropping afterwards; and US cocoa imports, which had been only one-half of France's in 1960, exceeded the latter in 1965.[1] Thus, despite French overall predominance in the economic and other spheres, the distinct US influence was being felt.

* * *

The above are the actual major US trade flows which have developed with states of divergent socio-political systems—from colonial in the South to progressive in the North, as well as with other countries in the various imperialist spheres—British, French, Belgian and the US. Their contours are distinctively political reflecting US imperialist aims and strategy, and at the same time are directly or indirectly bound up with American monopoly investments and profits, as well as aid.

These flows also are, in depth, the product of superimposition on longer-term US foreign economic policies, such as the advocacy of "free trade" abroad but protectionism at home, which have fostered US capitalist industrialization and expansion of the home market. The system of tariffs, quotas and other barriers, designed in the 19th century to protect American infant industries against the European powers, has contributed its share to the emergence of giant monopolies, which long ago have outstripped most of their rivals.

The US high technological level, size of market and overall economic might in industry and agriculture held an even more striking advantage over the underdeveloped countries' appendage economic structure, i.e., all of Africa (with the exception of South Africa). The disparity in strength and

---

[1] *Marchés Tropicaux et Méditerranées*, 3.IX. 1966.

resultant capitalist profit maximization by US commerce and industry (with the industrialists of South Africa, there is a profit-division based on African labor more comparable, perhaps, with the relationship between the monopolists of the United States and Western Europe) have had at best little regard for the effect on African development and have at worst directly hindered the latter.[1]

Today, the continuing protectionist and other economic policies of the United States in maximizing exports and minimizing certain imports have the effect of still further expanding the size of its market and of diversifying its products, i.e., of providing even greater economies of scale and lessening dependence on particular foreign commodities, and thereby improving its margin of advantage and bargaining position.

In seeking to rationalize the long outmoded need to shield infant industries, the present-day protectionist argument centers about defending established US domestic industries against the lower wages and costs of production abroad. On the other hand, when the underdeveloped country argues the need to overcome its obvious infant-industry and small-market plight, the industrialized capitalist countries urge upon them what would amount to a perpetuation of the status quo—international specialization along traditional lines of "comparative cost advantage". Not that the latter

---

1 Historically, the evidence is rather overwhelming. In four centuries of slave trade, the loss of population in the total slave trade to the "civilized" world "cost Negro Africa 100,000,000 souls" (W.E. Burghardt Du Bois, *The Negro,* N.Y., 1915, pp. 155-56). In the 17th and 18th centuries, the slave traffic was more profitable than trade in gold, ivory and pepper. The 19th century "traders' period" or "traders' frontier" witnessed a transition from a trade based on hunting or gathering (traditional societies and slave labor) to one based on agriculture and mineral production (wage labor). Urban growth, however, was not accompanied by industrial development, but constituted rather an evolution of trade centers, e.g., Leopoldville (palm oil, etc.), Accra (cocoa), Lagos (palm oil, cocoa, peanuts), Ibadan (cocoa), for the expansion of agricultural exports to Europe. The net result of such trade for Africa in general was stagnation and poverty, according to liberal writers like G. Myrdal, many Africanists and Marxists. Others who would like to save face for colonialism argue the advantages for Africa of increased monetary income, improved transportation, etc. (See S. D. Newmark, *Foreign Trade and Economic Development in Africa,* Stanford, 1964.) The decisive question even in the strictly economic context is how much further would African countries be today if they had been sovereign forward-looking states rather than under imperialist domination.

could not be, in many cases, of mutual advantage, but that commerce and other relationships in the capitalist world are guided by the principle of bargaining strength (economic, political, military) and as a result are reflected in detrimental price instability, worsened terms of trade, and dependence.

US trade with Africa in the 60's reveals a balance of trade surplus in general of $200 to $300 million annually (see Table IV). Politically motivated drops below this level—but still showing a $100 to $150 million favorable balance—were registered in 1961-62 (resulting from increased US imports from South Africa) and in 1968 (from, on the one hand, drastic cuts in US exports to the UAR and, on the other hand, a rise in tropical food and raw material imports from countries such as the Ivory Coast and Angola).

The general skewed trade, in both magnitude and composition, of the United States with most of Africa, which constitutes a handicap for the latter—either for increasing export earnings or for development—results, in large part, from the continuation of the general bias of US commercial and aid[1] policies. Tariffs and other barriers[2] of the industrialized capitalist countries are specifically regarded as preventing the trade expansion of the less developed countries by the General Agreement on Tariffs and Trade (GATT), e.g., as indicated in the Haberler Report (1958) and in the Programme of Action (May 1963).[3]

This applies to the entire range of US imports: to primary products—both non-competing (tropical crops) and com-

---

[1] US-tied aid in 1960-64, for example, led to the less developed countries' increase in imports from the United States one-third faster than the growth of their exports, while their imports from other industrialized capitalist countries increased only half as fast as corresponding exports.

[2] These range from import quotas and exchange controls to other, more sophisticated, non-tariff barriers, e.g., overevaluation for customs purposes, various administrative delays, and difficulties in "labelling for health purposes". See H.G. Johnson, *The World Economy at the Crossroads,* Montreal, 1965.

[3] The Common Market and Commonwealth through their blocs, and the United States and other industrialized capitalist countries through their individual policies which are not at all directed to facilitating "the efforts of less developed countries to diversify their economies, strengthen their export capacity, and increase their earnings from overseas sales" (Point 8 of the GATT Programme of Action). See *Economic Policies Toward the Less Developed Countries,* by H.G. Johnson, Brookings Institution, 1967.

peting (temperate agricultural crops and certain metals), and even more so to manufactured goods. The United States does not permit free access to its market even for most tropical products (as do, in contrast, Norway and Sweden), arguing first, that there would be no large expansion of the market since responsiveness of demand and supply to such commodities is small, and secondly, that a removal of duties would lead to US budgetary problems.[1] But since these arguments scarcely justify depriving the underdeveloped countries of a portion of their potential earnings, even US big business must at least voice the expediency of progressively eliminating such duties.[2]

In the case of a number of competing primary products (copper, cotton, iron ore, petroleum[3] and foodstuffs), US domestic subsidies and price supports have tended to reduce US imports. Moreover, US insistence on quantitative restrictions on a wide range of products of its highly mechanized agriculture and its inability to market its surpluses have led it to oppose stubbornly agricultural trade and commodity agreements within GATT, aimed at stabilizing prices and earnings. The double-edged US surplus disposal program (especially under PL 480 beginning in 1954), even if of immediate benefit to some receiving states, has tended to restrict exports by the underdeveloped countries by making receipt of such US commodities conditional on the recipient's restricting its exports of commodities in surplus supply in the United States, and also by replacing similar exports of other underdeveloped countries.[4] Removing the overall protection given to agriculture alone in the industrialized capitalist countries could mean an addition of one-sixth to the earnings of the less developed countries.[5]

---

[1] See *Trade Policy Toward Low-income Countries,* Committee for Economic Development, N.Y., June 1967.

[2] Ibid.

[3] The present oil quota system, for example, is favored by the oil monopolies over a proposed tariff, which could result in increased imports and lower domestic prices for petroleum and derivatives—perhaps cutting into profits, but saving American consumers between 4.5 and 7 billion dollars a year, according to Representative Charles Vanik (D.-Ohio). *Herald Tribune,* July 18-19, 1970.

[4] This involves an estimated loss of earnings of over two-thirds of a billion dollars annually, not to speak of depressing world prices for other producers, according to H.G. Johnson, op. cit.

[5] Ibid.

In the very small but critical area of manufactures, the typical pattern of US tariff protection is: The more the basic material has been processed, the higher the tariff. The bias against the infant industries of the less developed states is all the more striking when it is realized that only about 1 per cent of the total consumption of manufactures in the industrialized capitalist countries comes from the former—mainly from a few countries in Asia and Latin America but potentially from Africa as well. The elimination of tariff barriers on manufacturing imports from the less developed countries could increase US imports from them by about 50%.[1]

In capitalist world trade, the underdeveloped countries, not least of all Africa, are basically hampered by their unequal economic strength and bargaining position resulting from the weak and lopsided structural pattern of their economy and trade. This, in turn, continues largely as a consequence of their relations and ties with the imperialist countries and the latters' foreign economic policies. Although viewpoints on this question are as widely disparate as the many interests involved (e.g., foreign monopolies, national bourgeoisie, workers and other exploited classes),[2] the fundamental imbalance in trading strength is nevertheless generally recognized, e.g., by the United Nations and in the

---

[1] Estimates of Johnson and Balassa (ibid.).

[2] It is not surprising that a voice of US big business, the Committee for Economic Development, blames the underdeveloped countries for their plight arguing, in particular, that population growth is seriously retarding development, approaching "the feasible rates of increase in economic output, thus preventing significant growth in per capita income". (*Trade Policy Toward Low-income Countries.*) Those who see, and would help Washington meliorate, the conflict of interests, including academics and others close to the US government, advocate, e.g., "some constraint in the abuse by the powerful of their power over the weak" (H. G. Johnson, op. cit.) in the form of more liberal trade and development aid for political considerations. Representatives of national economic development like Raul Prebisch see the slow growth in underdeveloped countries' export earnings as essentially a consequence of technological progress, i.e., a product of structural factors (synthetics, lower demand for primary products) aggravated by trade barriers leading to a deterioration of terms of trade. They consider it an *obligation* of the industrialized capitalist states to transfer back income to the less developed, and to help them increase their export of manufactures through alteration of commercial policies, e.g., by granting preferences without reciprocity. Marxist-Leninists see the trade problem in the broader political-economic context of the class and national struggle against exploitation and imperialism.

rules of GATT. Africa's efforts to correct this imbalance, however, come into continual conflict with the imperialist powers—mainly through the Common Market and Commonwealth blocs, on the one hand, and the United States, on the other.

US foreign trade policies, which continued on the whole in the 1960's when most of Africa had achieved independence, were directed broadly at hindering their industrialization and increased export earnings (needed for development). Vivid evidence of this was the total complex of US positions taken at UNCTAD, Geneva, in March-June 1964. The United States was generally identified there as the least willing of the industrialized capitalist countries even to consider a "new" international division of labor which would permit the developing countries to industrialize.[1] Furthermore, the United States, was often the only opponent of the latters' demand for preferences to increase export earnings. In the final act of UNCTAD, the United States had the most negative votes against the demands of the underdeveloped states. Of the 15 general principles, the United States *alone* voted against principle one ("respect for the principle of sovereignty, equality of states, self-determination of peoples, and non-interference in the internal affairs of other countries"); *alone* against principle four (acceleration of growth and narrowing of income gap between the developed and less developed countries); *alone* against principle six (increased export earnings of less developed countries regardless of social system); and against principle twelve (disarmament-freed resources to be used for development). On other principles, the United States had the company of other imperialist powers in opposition to the underdeveloped states.[2]

The position of the Socialist states at the UNCTAD, it

---

[1] J. C. Mills, "Canada at UNCTAD", *The International Journal*, Vol. 20 (spring 1965).

[2] Thus, the United States also was against principle two (no discrimination on basis of socio-economic system); principle three (freedom to trade and dispose of own natural resources); seven (international arrangement for market access, remunerative prices for primary products); eight (concessions in preference to the less developed countries without demanding return concessions); eleven (increased aid without political or military strings). On special principles, the United States voted against principle one (setting targets for trade expansion); seven (compensation financing of worsening terms of trade); eight (surplus disposal by international rules); twelve (action to promote invisible earnings).

may be noted, had been in direct contrast. This, moreover, is more than borne out by their commercial policies in practice. Thus, Socialist trade, which is still of modest proportions for Africa as a whole,[1] nevertheless has supported a number of countries which have struck out determinedly for political and economic independence, e.g., the UAR, Ghana and the Sudan, which accounted for about three-fourths of the Soviet Union's trade with Africa from 1958-65. Indicative of such support was the trade structure and composition—imports comprised their major products, including manufactures, and about three-fourths of exports consisted of capital goods and equipment.

It is generally appreciated that Socialist trade, which has no export-of-capital or other capitalist drives for profits abroad, has supported its African trading partners economically through such guiding principles as bilateral balancing, purchases at world market prices or slightly higher, stable buying with increases in poor marketing years or at least not falling below levels of previous years, the provision of markets which either create trade for the African states or alternative outlets helping to support prices. The typical bourgeois criticism of Socialist trade is that it is also political, i.e., has the purpose of seeking to undermine Western ties—rather than possessing altruistic aims. But, if Socialist policies help to build a strong, independent Africa, which of itself resists the unequal terms of the neocolonialists, this is simply another confirmation of parallel interests.

Within the framework of the joint imperialist interest in keeping down the newly independent African states, the United States in its rivalry with the European powers has had to contend with their advantageous trade blocs and the lack of a big one of its own. The EEC with its overseas associated states and Britain with its Commonwealth have had duty-free or preferential access for their exports reciprocally in their respective blocs—which tend to continue the latters' traditional exports and imbalanced economy, fragmentation of bargaining power, and consequent dependence.[2]

---

[1] About 6%; about 80% of Africa's total trade is with industrialized capitalist countries.

[2] Most African states are dependent on either 1 or 2 commodities for over 75% of their total exports: 33 countries on one and 7 on two primary commodities. (Based on material presented at the 2nd UNCTAD, March 1968.) Although African trade is small (about 5% of world total),

This also excludes, or puts at a disadvantage in their markets, the United States (although not many of its overseas plants), as well as other non-bloc countries.

To be able to make full use of its economic strength, the United States in the early 60's pressed for the "open system" —all industrialized capitalist countries to have economic, commercial and political relations "without discrimination", i.e., to exploit on an equal footing. These efforts failed, however, because of the unwillingness of the European powers to abandon their advantageous commercial positions and of US sectional agricultural and manufacturing interests to drop US trade restrictions.[1]

Political-military considerations which had led Washington to support the EEC also led to US acceptance within GATT of common market treaty features of support prices for agricultural products above the world level coupled with levies on imports, and of the preferential arrangements with the former African colonies. The first involved an increase in protectionism abroad, and the second—new preferences in contravention of the GATT rules. But, Washington was looking hopefully to the Kennedy Round of negotiating reciprocal tariff reductions as the way to opening and linking up with the EEC (and, thereby, to their African bloc) markets. But this effort failed—at least temporarily—due in large part to the political conflict with de Gaulle, who spurned the Washington-endorsed British application to join the Common Market in 1963 (and again in 1967), plus the desire of other EEC members to prevent an aggrandizement of US influence.

Refusal of Washington to recognize the defeat of its commercial policies and strategy led to US virtual isolation at

---

it forms a higher proportion of national output (which makes for greater sensitivity to world prices and dependence on the big capitalist countries, who take about 9/10 of its exports) than that of industrialized countries, e.g., as much as 50% in several countries as compared to about 15% for such a big trading country as Britain. The former colonial power continues to be the main trading partner, e.g., few of the French-speaking states send less than 50% of their exports to France, and Senegal up to 86%; imports from France are equally high—up to 66% in the case of the Ivory Coast and 68% in the case of Mauritania in the mid-sixties.

[1] See "The Future of the U.S. Foreign Trade Policy". Hearings, Committee on Foreign Affairs, Subcommittee on Foreign Economic Policy, July 1967.

the UNCTAD in 1964. Its advocacy, from positions of strength, of free trade, "non-discrimination" and "reciprocity" came into headlong collision with the demands of all the underdeveloped states for higher prices for primary products, international commodity agreements, modification of protectionist policies which restrict their market, and particularly one-way preferences. The US delegate stood apart even from his West European colleagues, although the latter also were unwilling to extend preferences or the most-favored-nation principle to underdeveloped states—other, of course, than those in their trade blocs. The US opposition to preferences continued unaltered throughout the 1966 meetings of the UNCTAD preferences committee.

By 1967-68, realizing that Washington's efforts to phase out the preferential system had failed and that the problem for US monopolies might even be aggravated if the European Common Market enlarged its scope (and Britain joined), representatives of US big business circles were perplexed as to how to alter commercial strategy. In testimony to Congress, for example, former Under Secretary of State Ball was completely discouraged over the prospects of achieving the "open system",[1] although David Rockefeller of Chase Manhattan Bank was not and urged Washington to keep trying. The consensus of big business, however, was that preferences were in the cards politically and if the big capitalist states wished to retain their influence in the underdeveloped states some steps would have to be taken.[2]

This did not imply that, however much the United States stood for "free trade" for its own exports, US protectionist walls would fall like those of Jericho, or that US monopolies were prepared to make great sacrifices in permitting imports from the less developed states. In the first place, there were many rows of other US trade barriers to fall back upon, such as imposed or voluntary quantitative restrictions,[3]

---

[1] So much so that he felt that it might be well to recognize European primacy in Africa, as the United States enjoyed in Latin America, ibid.

[2] Committee for Economic Development, op. cit., and *Issues and Objectives of U.S. Foreign Trade Policy,* U.S. Congress, Joint Economic Session, Washington, September 1967.

[3] The 5-year cotton textile arrangement of 1962, for example, was regarded by some exporting countries as bordering on fraud (G. Patterson, *Discrimination in International Trade: The Policy Issues, 1964-65,* Princeton, 1966). US officials reporting on a renewal of this arrange-

which could be nominal or rigorously enforced, or red tape and collusion. Secondly, it was felt that in any case the net effect on the US economy would be small, and, thirdly, minor preference concessions might turn the less developed countries away from import substitution (i.e., restructuring their economies) and to the world capitalist market without feeling victimized by discrimination. This would help continue the same merry-go-round of lopsided structure and dependence—either on European or American imperialism, or some combination of both.

Thus, by the 70's it appeared that Washington was embarked on a new sophisticated tactic in foreign trade policy whose fundamental motivation was to gain political credit in the underdeveloped countries with such minimal concessions as dropping its opposition to their demand for preferences. This essentially paralleled the US big business recommendations of the late 60's. At the same time, the United States would continue to push for reversal of "reverse preferences" which favored its rival European imperialist powers. In this connection the Nixon Administration offered the African countries not to tie US aid to purchases from the United States in return for entry into their markets.[1] The game was still the same, although the tactics were changing.

The overriding contradiction between imperialist states, despite their rivalry, was more broadly corroborated at the Third UNCTAD at Santiago in April-May 1972 when the United States and other industrialized capitalist countries refused no less categorically than at the two preceding conferences to make concessions to ease the plight of the developing countries. Thus, they turned down such demands as using partial arms reduction expenditures for the benefit of the developing states, lightening their debt burdens, or permitting their participation together with the IMF and GATT in a permanent committee on currency questions. (The last proposal was advanced in the light of the serious losses caused to their trade position by the imperialist countries' currency revaluations in August 1971.)

---

ment for a 3-year period beginning in 1967 stressed the expectation of slowing down rather than increasing imports. (*Issues and Objectives of U.S. Foreign Trade Policy*, op. cit.)

[1] *U.S. and Africa in the 70's*, Washington, D.C., April 1970.

3. SOCIAL STRUCTURE AND RELATIONS

AFRICAN SOCIAL FORCES AND THE UNITED STATES

To better understand the inter-relationship of the United States and African social forces, it is necessary to begin by examining briefly the development of the social formations on this continent. Like its predecessors, US imperialism generally had been able to take advantage of and by the same token has strived to perpetuate the backwardness of social orders. However, as these societies have slowly developed—in response to inner impulses as well as to foreign exploitation—forces have emerged which more and more have begun to challenge the status quo. Of major importance in this connection has been the powerful impetus provided by the example and support of the Socialist countries.

The tendency of bourgeois writing to ascribe ethnic, racial and similar reasons for the retardation of "Dark" Africa's socio-economic development has lost its appeal in the post-war period largely due to the collapse of colonial empires, the appearance of numerous newly independent countries in the international arena, the remarkable progress made under Socialism in a brief historical period by such previously backward regions as Mongolia,[1] the Central Asian Soviet republics,[2] China, North Korea and North Vietnam. It is not surprising therefore to find more recent Western apologetic shifting its emphasis to Africa's natural and demographic conditions.

Africa, which in pre-colonial times has known highly advanced civilizations (Songhai, Mali, Oyo, Benin, Ghana and Zimbabwe), is not without serious natural difficulties

---

[1] For two centuries before the revolution, Mongolia was a neglected colonial hinterland dominated by local feudal lords, foreign trade and usurer capital. It lacked industry, modern transport and agriculture, and even a monetary system of its own. In contrast, from 1940 to 1964, gross output increased ten-fold, with industry accounting for more than 40% of gross national product. Annual rate of growth averaged 10.3% in the 60's. Today, workers constitute more than one-third of the population. See Y. Tsedenbal, "The Revolutionary Party and Social Changes" in *World Marxist Review,* February 1966.

[2] With all-Soviet aid, industry rapidly grew (as a percentage of GNP) from 1928 to 1932: Kazakhstan—17% to 44%; Uzbekistan—30% to 53%; Tajikistan—23% to 44%. Мировая социалистическая система хозяйства *(World Socialist Economy)* in four volumes, Vol. 1, M. 1966.

affecting progress, e.g., a frequent lack of fertile and cultivable land,[1] inadequate or excessive rainfall and water supply creating deserts or impenetrable forests, or the wide prevalence of malaria and the tsetse fly which rules out cattle-raising and infects human beings with the deadly sleeping sickness. In tropical Africa particularly, geography and climate have no doubt acted as a brake, with the agricultural work season in the savannahs only 100-150 days, in the equatorial forest regions 60-80 days, and a dry season which is "dead".

Without minimizing the adverse influence of such factors, however, they alone can scarcely explain the failure under colonialism to promote a transition from the wooden mattock to better implements, from helplessness in the dry season to some forms of irrigation, and from handicrafts to at least light industry.[2] Social development, too, has remained retarded at the traditional, pre-class or elementary feudal level in most of the continent, with such notable exceptions as Egypt, Maghreb and northern Sudan where patriarchal society had been largely replaced by class formation and an emergent ethnic nationality before the advent of the European powers.

The vestiges of pre-colonial social structures, which were preserved and upon which were superimposed colonial changes, add up to a complex mosaic which no single model can hope to explain.

Lack of progress links most closely with economics, and this was dependent on relations with and the socio-economic policies of the colonial powers. In the period until the mid-nineteenth century, when the slavers' trade in men was dominant and trade in goods negligible, tribe was used against tribe in the hunt for slaves. Thus, in almost four centuries during which countries like Britain and France progressed from feudalism to capitalist industrialization and nationhood, the export of 50 to 100 million African slaves, mainly to America, ruined villages, drained manpower and accentuated

---

[1] Thus, for example, from 8% in Libya and 20% in Algeria, to 48% in Morocco; the 3% in the ARE limited to the small but rich Nile valley.

[2] In 1958, Africa (minus South Africa) accounted for only 0.7% of the light industry of the world capitalist economy, 90% of which fell to five countries. V. V. Rymalov, *Disintegration of the Colonial System and the World Capitalist Economy*, Moscow, 1966, pp. 300-304.

tribal division and strife. Economic plunder through the system of feudal and tribal chiefs, under the wing of colonial authorities,[1] made use of tribal customs and feudal and semi-feudal forms of exploitation, e.g., tithes and forced labor, and thereby tended to perpetuate social and economic stagnation.

With the transition to monopoly capitalism in the last quarter of the nineteenth century, imperialist export of capital sought to expand the profitable output of African raw materials by supplementing those derived from earlier forest gathering (based on traditional societies) with a more regular and intensive exploitation of African lands and mines for the "workshop" of Europe. By the first decade of this century, the readily accessible areas of tropical wealth (e.g., rubber, timber and wild palms in French Equatorial Africa and the Belgian Congo) had been drained[2] and exports began to decline. A deeper penetration and widening of the market was called for and could be provided with the assistance of modern transport, e.g., for groundnuts in northern Nigeria and Senegal.

Trade then was supplemented by productive functions. In West Africa, the introduction of new perennial crops, e.g., cocoa, coffee and rubber, for the market led from the production of annual crops on communal land to African investment in land, inheritance of such investment, and the emergence of a class of small farmers and traders. In East, Central and southern Africa where the climate was most favorable, large numbers of European settlers had expropriated the best lands[3] for plantations, farms and mineral

---

[1] The misuse of chiefs and feudal rulers in "indirect rule" was widespread, e.g., in Northern Nigeria, the chiefs of Ashanti in Ghana and in Uganda. Regardless of whether the chiefs' authority derived from their traditional position or from their role as agents of colonialism, writes Professor L. Delavignette, formerly Gouverneur Général de la France d'Outre-Mer, the French colonial administration would have been "helpless" without the traditional chiefs, and use was made of them from the beginning. See *Colonialism in Africa, 1870-1960,* Vol. II; *The History and Politics of Colonialism 1914-60,* ed. by L.H. Gann and P. Duignan, Cambridge University, 1970.

[2] See W. F. Barber, "The Movement into the World Economy" in *Economic Transition in Africa,* ed. by M.J. Herskovitz, London, 1964, p. 301.

[3] The percentages varied from 7% in Kenya, 9% in the Belgian Congo, 49% in Southern Rhodesia and Swaziland, to 89% in the Union of Sounth Africa. M. Hailey, *An African Survey Revised, 1956,* p. 689,

exploitation for domestic and foreign markets which required and forced hundreds of thousands confined to the poor soils to become wage laborers.

Thus, on a continental scale and in a great variety of forms, a dual economy was created comprising a sector geared to export, with class differentiation emerging from and existing side by side with a stunted natural economy with its traditional social structure. The undermining of old economic relationships, however, did not witness the evolution of a national market but rather an increased suction of raw materials from mines and plantations through railroads and ports—a series of enclaves owned or dominated by European trading companies in West Africa, or foreign plus settler mine-owners, planters and farmers to the east and south.

Delayed and distorted social development has centered largely about the level of the means of production and the degree and forms of exploitation of labor. The draining off abroad through expatriation of profits, interest on debt, and colonial remittances of the surplus product and even part of the necessary labor required to reproduce the African worker—his miserable wages and conditions—has all but precluded local accumulation and has been an obstacle to technical progress. In West Africa, the "banana motor", as the Ivory Coast plantation owners refer to the Negro worker receiving little or no pay and nourished on bananas, costs less than a machine or even elementary tools.[1] A cheap pool of untrained intensively exploited laborers who know only the most simple tools or none at all[2] is needed.

To obtain such labor, external force and compulsion in various degrees have been used at different times—slavery, land seizures and taxation to break up subsistence farming and drive the African to work in the money-commodity economy, and the use of chiefs to recruit their people. This

---

London, 1957. See J. Woddis, *Africa—the Roots of Revolt,* Chapter I, London, 1960.

[1] Jean Suret-Canal, *Afrique Noire, Occidentale et Centrale, l'Ere Coloniale (1900-45),* Paris, 1964, pp. 90-91.

[2] As late as the 1950's, in the Gold Coast, thousands of men have never seen a pick or shovel; in Nigeria, laborers carry baskets of ore on their heads; in the Ivory Coast quarries, men work without even the shovel and wheelbarrow because "it wasn't worth while to teach them as they were engaged only for a few days or weeks". Sources cited in Woddis, op. cit., pp. 152-53.

was at such variance with the norms even in capitalist Europe that world hostility to the worst of these practices influenced the major colonial powers (except Portugal) in 1930 to sign the Geneva Convention on Forced Labor. This, as indicated earlier, also had its repercussions on forced labor in the US area of "special responsibility"—Liberia.

Such practices, nevertheless, have continued either through exceptions to the Convention or in related forms. Direct forced labor varied from services for native chiefs to corvée (obligatory work for public services). The head or hut tax payable only in cash, which compelled Africans to seek employment for wages, e.g., in the Congo, Ruanda-Urundi and Tanganyika, led to the concept of "target worker"—to get money for taxes and the bride price, commuted from cattle to money.

In southern Africa, figures of forced migrant labor have run into hundreds of thousands,[1] particularly of young men leaving their villages to work in the mines and on European farms. Portuguese colonial authorities under long-standing agreements continue to provide well over 100,000 contract workers a year to South Africa,[2] some two-thirds of whose African miners come from other territories.[3] Restricted to labor camps, prevented from shifting to other urban employ-

---

[1] *African Labour Survey,* International Labor Office, Geneva, 1958, pp. 137-44, Hailey (revised), op. cit., pp. 1377-79.

[2] The official number of Mozambique migrant workers in the early 60's was: 169,000 in South Africa and 187,000 in Rhodesia. For the historic roots and economic impact of this phenomenon see В. Л. Шейнис, *Португальский империализм в Африке после Второй мировой войны (Portuguese Imperialism in Africa after the Second World War),* M. 1969, pp. 165-82. Reliable estimates of Mozambique permanent migrant workers amounted in the late 1950's to 500-600,000 (official statistics 360,000 to 400,000), or 2 adult males out of every 5 or 6. (An estimated 100,000 Angolans also were working in South Africa.)

[3] The process began when English colonists, who took over the Cape Colony in 1806, began ruthlessly to burn and slaughter, confiscate cattle and land, and convert the tribal people into hired laborers and customers for English goods. With the discovery of diamonds in Kimberly in 1867 and gold in the Witwatersrand in 1886, processes necessary for capitalist development were speeded up—expropriation of African land, poll taxes, hut and animal taxes, pass laws and labor control. The gold mines could not get enough labor and immigration (including Chinese workers for a short while) was stepped up. At the request of the gold mine owners, a Government Commission in 1903 recommended modifying the Native Land Tenure System to force the Africans to work in the mines by alienating them from the land. The 1913 Land Act,

ment, easily displaced by other migrants, unskilled migratory labor is tied to low wage scales generally based on minimal subsistence for the bachelor worker, with the rationalization that the worker gets additional support from his claims upon his native village[1] (where his family lives in tribal or communal society and to which he returns after one or two years).

Such capitalist exploitation, which is incorporated in the rigorous system of racial and social discrimination, is shared in through investment and trade by the big monopolies of the imperialist powers, including the United States, making them at least silent partners in apartheid.

Thus, in contrast to Europe, where class formation—especially bourgeois and proletarian—developed and matured from largely internal forces in the economic sphere, a nation was formed in connection with the more advanced capitalist industrial process, and the bourgeoisie's bid for power represented a political-economic struggle against a backward feudal land-owning aristocracy, Africa presents a more complex dynamic. Here, external imperialist forces have played a big, if not decisive, part. This is not to discount such internal countervailing or other forces as: in North Africa, a strong Arab national, ethnic and religious movement; in South Africa, the largest concentration of European settlers, who have evolved an industrialized colonial society of their own; in tropical Africa, emergent social forces, who are not indifferent to world events and ideas.

The outlived or stunted social formations inherited at independence[2] reveal a varied composite of dominant but

---

which made it illegal for Africans to occupy land outside the "native reserves" (now called Bantu homelands) comprising less than 13% of the country, was made possible by Britain's handing over political power to a privileged white minority of the four colonies—Cape, Natal, Transvaal and the Orange Free State—in the South Africa Act of 1910. See Duma Nokwe, *The National-Liberation Movement of South Africa,* paper presented at the Scientific Congress against Racism and Neocolonialism held in Berlin in May 1968.

[1] *African Labour Survey,* pp. 147-60; see also W.E. Moore, "Adaptation of African Labor Systems to Social Change", Chapter 13, in *Economic Transitions in Africa.*

[2] For broad insights into and detailed treatment of socio-economic development in African and other underdeveloped countries see, for example, *Классы и классовая борьба в развивающихся странах* (*Classes and Class Struggle in the Developing Countries*), in three volumes, ed. by V. L. Tyagunenko, 1968, Vol. III, especially Chapter V;

weak classes and strata, with resulting coalitions of parties in power, e.g., of feudal-landowners, and of various capitalist tendencies. In North Africa (alone) a national bourgeoisie had emerged before independence as a definitely formed class in the Maghreb, Egypt and Sudan. Nevertheless, feudal elements remain relatively strong, and also represent a force, despite agricultural reforms, in the ARE and Algeria (as well as in central Ghana, north and west Nigeria, Uganda, Zambia, Ruanda, Burundi; and still predominate in Ethiopia).

In tropical Africa, where there is an interweaving of patriarchal-feudal, dying tribal and developing capitalist relationships, the national bourgeoisie is extremely limited since industrial production, wholesale, and to some extent even retail trade are usually in the hands of Europeans (or of Lebanese and Syrians in West Africa; Indians, Greeks and Armenians in East Africa).

Instead, an African middle or petty bourgeoisie is to be found mainly in retail trade, or as middlemen between peasants producing export crops and foreign trading companies. On the land, a bourgeoisie has emerged from farmers growing cash crops for export, notably in Ghana (cocoa), Senegal (groundnuts), the Ivory Coast[1] (coffee, cocoa), e.g., President Houphouet-Boigny is a large landowner, Liberia (rubber, coffee and cocoa), Dahomey (palm nuts) and Cameroun (coffee). Moreover, this stratum is able to branch out into domestic commerce and transport.[2]

A part of the bourgeoisie, with ties to imperialism, does not invest its capital domestically but together with highly paid civil servants and officials (e.g., in Liberia, Kenya, Ni-

---

V. V. Rymalov, op. cit.; И. И. Потехин, *Нации и национальный вопрос* ("Nations and the National Question") in *Africa, An Encyclopaedic Handbook,* Moscow, 1963; *Анти-империалистическая революция в Африке (Anti-imperialist Revolution in Africa),* ed. by V. G. Solodovnikov, M., 1967, Chapter I.

[1] The one-sided expansion of the export sector tied to foreign economies and the emergence of a bourgeoisie of some 20,000 plantation owners, has led to growth but without development, with foreigners receiving about 40% of the income in the productive sectors and holding all the key positions. Remittances abroad in 1965 amounted to 25.2 thousand million African francs equal to twice the amount of aid plus private capital inflows. S. Amin, *Le Développement au Côte d'Ivoire,* Paris, cited in *West Africa,* December 26, 1967.

[2] See *Raymond Barbé, Les Classes Sociales en Afrique Noire,* Paris, 1964; J. Woddis, "African Capitalism", in *Marxism Today,* May 1966.

geria, Uganda and Ghana) prefers to acquire shares in foreign companies.[1] The new "bureaucratic" bourgeoisie makes use of its disproportionate voice in state power to acquire personal wealth, privilege and luxury not in keeping with the modest resources of the country.[2]

In view of the relative weakness of class development, the African intelligentsia, which is of uneven social origin, plays a particularly important role. Trained in administration, education, medicine and military affairs, mainly for service under colonialism, many have acted as an élite in behalf of older ruling classes. Others have sought to balance between older and newer class forces, both domestically and internationally. A section has cast its lot with national interests and Socialist ideas—leaders such as Lumumba, Touré, Keita, Nyerere, Nkrumah—leaning for support on the young working class and broad peasant masses.

Although imperialism has relied in the colonial past on the old aristocracies, feudal landowners and traditional chiefs, nevertheless, the struggle sometimes has cut across these lines, e.g., feudal leaders in Morocco who took part in the anti-colonial struggle in the mid-1920's; the Ivory Coast plantation owners, who helped eliminate forced labor in 1946 to have more available wage labor, constituted a bourgeois-feudal stratum of influence in the independence movement; chiefs in South Africa in 1912, who were, together with middle class intellectuals, the founders of the African National Congress, included its former President-General, Chief A. Luthuli.

In the period of post-independence for most of Africa, US social policy, paralleling American political aims and monopoly interests, continues to rely heavily on the dominant classes of outlived socio-economic formations and those with former colonial ties to imperialism. As part of US political, economic and military relations, this is the most convenient and long considered the cheapest avenue of exerting influence and maintaining social "stability". Thus, feudal land-

---

[1] T. Geiger and W. Armstrong, *The Development of African Private Enterprise,* Washington, 1964, pp. 77-78.

[2] Thus, in the 1964-65 Senegal Budget, according to a study by G. Chaliand, 47% of total expenditures was under the head of "personnel"; in the Ivory Coast, for a group representing 0.5% of the total population, it was 58%; in Dahomey—65%. *L'Afrique dans l'Epreuve (Africa on Trial),* a symposium, ed. F. Maspero, Paris, 1966.

owners and comprador bourgeoisie in several of the northern Arab states; tribal and feudal leaders in tropical Africa; racist plantation owners in the Portuguese colonies, and a white capitalist class in South Africa. (See the following section "US Partnership in Social Oppression".)

In addition to major emphasis on such ruling classes, however, Washington has shown an active interest in new and developing classes, such as the small African land, commercial and bureaucratic bourgeoisie. This has been particularly marked in certain countries of tropical Africa where this new élite has an important share in state control.

In Liberia, for example, a US-encouraged bourgeoisie has developed, particularly since 1944, when President Tubman came to power, from Americo-Liberian descendants. Their main occupation is government position but they are also planters, owners of land and housing, and administrators of mixed companies in which the government owns half the shares. Most of the leading figures of the country are owners of plantations, which are on a much smaller scale than those owned by American monopolies, but with a total combined output of the same magnitude as the latter and with one of the lowest minimum wage scales—8 cents an hour.[1] Especially through production, marketing and other economic and financial ties, this bourgeoisie is tied to US big business with the common interest of suppressing and exploiting the working people.[2] After President Tubman died in 1971, a fortune of well over $100 million reportedly was left to his widow, and his elder son's father-in-law, William Tolbert, assumed the presidency.

In Kenya, British imperialism since the early 1950's has encouraged a small African capitalist farmer class, as well as a bureaucratic bourgeoisie,[3] which became no less an object of US attention and influence. Officials, including Cabinet Ministers, have acquired settler farms and succumbed to the

---

[1] *West Africa,* January 27, 1968.

[2] A recent attempt to mollify social unrest and opposition has been Tubman's "Unification Policy" of integration of the Americo-Liberians and the tribal peoples, which permits the latter "to identify with the Liberian nation through the personality of the President". See J. G. Liebenow, *Liberia,* Cornell University, 1969.

[3] The civil service after independence was still filled with expatriates, who tend to resist change. J. Oginga Odinga, *Not Yet Uhuru: An Autobiography,* New York, 1967, p. 247.

enticements of wealth.[1] The interests of the new élite has been voiced by such men as Kiano, Moi and Tom Mboya. Since 1964-65, this has found its political expression in a move to the Right within the ruling party KANU led by Mboya, Ngala and Kenyatta which resulted in the ejection of the Left from office and driving it from KANU.

The party conference at Limuru in March 1966 revealed the part played by US imperialism. Former US ambassador William Attwood in his book of memoirs[2] makes clear how he helped in the ousting of the "Reds". The sudden appearance of large sums helped the Right to intimidate and to bribe, to exclude "unfavorable" delegates from the conference, to reverse the election of Kaggia as KANU Vice-President for Central Province, and to strip Odinga of his post as Vice-President of the party.

Perhaps sensing that the African working class, although still young and undeveloped, is destined to play an ever increasing role, the imperialist powers have done their utmost to prevent its growth and organization. Nevertheless, as part of the historical process of extending capitalist exploitation to the mines and on the land, the number of wage laborers has increased—although most of them are seasonal and migratory. Of 18-19 million wage laborers in Africa, about 10-15% (not more than three million) may be considered strictly proletariat, i.e., dependent solely on wages for their subsistence.[3] The numerically small proletariat, correlating with industrial development and urbanization (miners, dockers, railwaymen), is predominantly in North Africa, followed by South Africa, then Zaïre and tropical Africa.[4] This is also reflected in roughly similar proportions of workers organized in the trade unions.[5]

---

[1] *African Communist*, No. 32, 1968, p. 11.

[2] W. Attwood, *The Reds and the Blacks*, New York, 1967. The succeeding US ambassador, Glenn E. Ferguson, had to disown the book as a "violation of ethics" and promised not to write about his activities "for five years after I leave Kenya". A. Lerumo, "New Light on Kenya" in *African Communist*, No. 30, 1967.

[3] *Рабочий класс Африки (The Working Class of Africa)*, ed. by I. P. Yastrebova, M., 1966, pp. 17, 29.

[4] Wage earners range from about 4% of the population in Nigeria and former French West Africa to 25% in Zaïre. Joan Davis, *African Trade Unions*, London, 1966, p. 24.

[5] Thus, of the approximately five million total membership in the mid-sixties, about one-half were in North Africa. For country breakdown, see *World Marxist Review*, February 1966, pp. 40-43.

Historically the African unions, early recognized as an integral part of the national-liberation movement, were banned in most countries by the colonial powers up until World War II. The right to organize was won mainly during the war, and in the independence movements of the postwar period union membership grew to about three million by 1960. Even though small, African unions represented a force not least of all in that they were closely associated with and even constituted sections of the trade unions of the metropolitan countries, especially France, Belgium and Italy. Similarly, the Communist movement in Africa, as the political expression of the working class, also developed with close ties to Europe in the postwar national-liberation struggle, mainly in the Maghreb and South Africa, and more recently in Senegal, Nigeria, Basutoland. Although membership on the continent is small, it rose from 5,000 in 1939 to 20,000 (1957), 40,000 (1961) and to 60,000 in 1967 in the face of great difficulties.

The growing trade union strength, however, was systematically undermined in certain national labor unions by imperialist and colonialist policies,[1] or as a result of the international splitting tactics of Right-wing labor leaders[2] in asso-

---

[1] In South Africa, for example, the ruling class-fostered social disease of racism seeped into and weakened the organizations of white workers by alienating them from the great African majority (two-thirds of the mine labor force comes from outside South Africa, i.e., is migratory, with one-two year turnover, and prevented from organizing, etc.). Similarly, the weapon of "anti-Communism": Thus, African trade unionism, which reached a quarter of a million in the 1920's, with members as far afield as Zimbabwe, Zambia and Malawi, declined after the expulsion of the "Reds"—the most hard-working and militant members. See D. Nokwe, op. cit.

[2] In 1947-48, coinciding with the launching of the cold war, the reformist leadership of the French Force Ouvrière and International Federation of Christian Trade Unions, the British Trade Unions Congress and the Belgian General Confederation of Labor, split the international trade union movement and fragmented African union organizations. Although the AATUF, founded on May 25, 1961 at Casablanca, restored unity to over four-fifths of Africa's trade union membership, nevertheless, the influence of the ICFTU is notable, for example, in Liberia, Sierra Leone, Tunisia, Nigeria, Ethiopia, as well as in former French West Africa. Malagasy, former French Equatorial Africa and the Congo (K) show the influence of the Christian movement (ICCTU). Kampala has been used as a Western center, and the Histadrut has been active in Ghana and Kenya. J. Meynaud and A. S. Bey, *Trade Unionism in Africa,* London, 1967, pp. 85, 92.

ciation with, or in the interests of, their respective capitalist classes—not least of all of the United States.

On the basis of a long history of cooperation with Right-wing trade union leaders, the US government[1] did not find it difficult to turn similar efforts to Africa in the postwar period. This has been done largely through the Right leadership of the merged AFL-CIO (in 1955) under George Meany and his appointed Director of the Department of International Affairs, Jay Lovestone, a professional anti-Communist and CIA man.[2] Emphasis has been on curbing militant African trade unionism, splitting it and cutting its international ties under cold war slogans.

This followed in the wake of the February-March 1957 African tour of the then Vice-President Nixon, who urged a broadening of American activities and particulary that US embassies become better geared to promoting US private investment and to curbing "Communist" (read: militant labor) activity.

"In every instance," he reported, "I made it a point to talk to leading labor leaders of the country I visited. I was encouraged to find the free trade union movement is making great advances in Africa ... have recognized the importance of providing alternatives to Communist-dominated unions and ... keeping the Communists from gaining a foothold ... I wish to pay tribute to the effective support given by the trade unions in the U.S. to the free trade unions of the countries I visited."[3]

To implement Nixon's recommended policies, Joseph Satterthwaite, then Assistant Secretary of State for African Affairs, in a conference of American Ambassadors and senior officers held in Lourenço Marques, Mozambique, reportedly urged taking advantage of the anti-colonial feeling against the European powers even though difficult officially because of the NATO partnership. One of the ways this could be

---

[1] This goes back to World War I, when Samuel Gompers created a Pan-American movement with funds supplied by President Wilson. By the early 1960's, the US government was spending over $13,000,000 a year on international labor affairs, and the AFL-CIO was devoting 8% of its budget to international activities. See J. Davis, op. cit., p. 201.

[2] See Sydney Lens, "Lovestone Democracy", in the *Nation*, July 5, 1965.

[3] *The New York Times*, April 7, 1957.

done, however, he advised, "is through the AFL-CIO contacts in the African labor movement".[1]

In the ICFTU, American delegates both vied for position with the British (over one-half of the African labor movement was from the British colonies) and clashed with them over policy. With its greater maneuverability, the US tactic was to seek to ride with the tide, to identify with the liberation movement and to support national trade union federations[2]—this, in sharp conflict with the British view.

The critical issue for the American labor leaders, however, was the ICFTU's insufficient cold war ideology. They differed with the British, for example, on making acceptable political ideology (i.e., anti-Communism) a prerequisite for international trade union assistance to Nigeria. In 1964, dissatisfaction with the ICFTU's lack of anti-Communist activity[3] also played a part in the AFL-CIO's cutting its financial contribution to the ICFTU's International Solidarity Fund, and then to its forming an autonomous AFL-CIO organization for activities in Africa.

The establishment of the African-American Labor Center in 1965 (AALC), with its director Irving Brown, a longtime associate of Jay Lovestone, also followed the failure after years of effort to pressure the ICFTU into accepting the former as secretary in charge of African affairs. The Center, modeled after, but smaller than, the American Institute for Free Labor Development in Latin America created in 1962,[4] also works in close cooperation with US government agencies, particularly the Agency for International Development (AID), to which it submits its projects for approval and

---

[1] British *Cabinet Paper on Policy in Africa,* December 12, 1959 (summarizing the December Brussels ICFTU conference), published by the Trade Union Congress of Nigeria, quoted by George Morris in *The Worker,* February 5 and 12, 1961.

[2] J. Davis, op. cit., pp. 192-96.

[3] The AFL-CIO attacked it as inefficient (their money allegedly was lying unused in banks). Ibid., p. 207.

[4] Since 1962, this Institute received some $21 million, publicly acknowledged, according to the Morse Subcommittee of the Senate Foreign Relations Committee (*Congressional Record,* September 25, 1968; *Daily World,* December 12, 1968).

The AALC, according to AID, had received in three years some $2.4 million from U.S. Federal funds and "another $1 million" from special funds set aside for special activities by U.S. ambassadors in African countries. (*Los Angeles Times,* May 22, 1966; *Daily World,* December 13, 1968.)

funds: vocational training, cooperatives, workers' education, health clinics and housing. There is little AID supervision of projects, because the US government "does not want to be too closely associated with them in public"[1] since the propriety of such a relationship might be questioned.[2]

Since 1965, the AALC has sent materials and equipment to the trade unions of Liberia and Senegal. Very large sums go to its Trade Union Institute for Social and Economic Development in Nigeria, and over 70 leading trade union officials had completed its courses by the close of 1967. In Kinshasa, a trade union cadre institute was set up for middle-ranking and junior trade union officials, from which 240 officials had completed instruction by the close of 1967. Funds also have been supplied to Sierra Leone.

Many of the African-American Labor Center's 34 projects in 16 African countries[3] (e.g., Kenya Tailoring Institute) are hardly designed to make a great impact on development, but are more often an excuse or cover to place agents and make contacts in African labor groups. As was revealed in the 1967 exposure of dummy foundations to channel funds through labor, student, research and cultural organizations, CIA funds have gone through the Baird Foundation to the African-American Institute and the American Friends of the Middle East; through the J. Frederick Brown Foundation to the American Society of African Culture.[4] One organization, Peace With Freedom Inc., also named in the CIA exposure, was headed by a Hungarian emigré, Robert T. Gabor, who was eventually ousted by the Kenya government for the CIA connections.[5]

As a result of undermining activities, American Right-wing labor leaders have become increasingly suspect in Africa. In 1962 the Nigerian Trade Union Congress objected

---

1 Ibid.

2 When the Senate Foreign Relations Committee called George Meany to testify in July 1969, Senator Fulbright asked him whether, as director of the African-American Labor Center, he considered it proper that 90% of its funds came from the State Department. *Daily World*, August I, 1969.

3 *Daily World*, December 14, 1968.

4 *The Worker*, February 26, 1967.

5 Irving Brown's own CIA ties were revealed by Thomas Braden, a former special assistant to CIA chief Allen W. Dulles, in an article entitled "I'm Glad the CIA Is 'Immoral'" in the *Saturday Evening Post*, May 20, 1967.

to the "harm being done by a battery of American spies in the guise of labor leaders in their efforts to sabotage a most desirable unity among Nigerian workers".[1] An appeal was also made at that time to declare Irving Brown a prohibited immigrant. Similarly, US Ambassador to Kenya, Glenn E. Ferguson, reportedly advised Vice-President Hubert Humphrey not to include Brown in his entourage when visiting Africa in January 1968 because there was no welcome mat out for him in Africa's unions.[2] The late Tom Mboya, who headed Kenya's unions before becoming Minister of Economic Planning, was said to have earned distrust because of his close relations with such American labor officialdom.[3]

In the battle against US imperialism and reactionary labor leadership, however, Africa's trade unions have not permitted themselves to be isolated or to be divorced from world issues. Thus, at a meeting in Prague, in May 1968, the Secretaries of the WFTU and AATUF decided to make better use of their strength as an international force by preparing a joint Consultative Conference of African and European Trade Unions. The resultant meeting in Conakry in March 1969 not only drew up a joint declaration, which censured US imperialism, the annexations by Israel—a base of imperialism, and racism in South Africa, Southern Rodesia and South West Africa, but also prepared the grounds for greater cooperation and unity of action.[4]

US PARTNERSHIP IN SOCIAL OPPRESSION

Southern Africa provides the most vivid example of US relations and policies to dominant classes. The orientation of Washington to the small ruling white minorities in southern Africa (about 5% in Rhodesia, 1% in Mozambique, 4% in Angola, 19% in South Africa)[5] of necessity involves the United States in the system of racism and colonialism.

---

[1] W. H. Friedland and D. Nelkin, "Differences and Policies Toward Africa", in *Africa Today*, December 1966.

[2] *Daily World*, December 13, 1968.

[3] *Los Angeles Times*, October 10, 1968.

[4] *Report of Activity of the World Federation of Trade Unions*, May 1965-April 1969, Budapest, October 1969, pp. 485-87.

[5] Based on UN statistics (rounded). Population data for Africa are not precise and should be interpreted as being only approximations. (See Table VI, p. 130.)

It would be difficult to overestimate the effect of racism on the African—not unlike that on the Black in the United States—where class structure is based upon or closely correlated with it. Thus, skin color is used to predetermine political rights, wages, education, housing and social status. The Europeans are far and away on top of the scale, the Asians much below them, and the Africans on the bottom. For the Africans, minority rule and colonialism—"something a white nation does to a darker people"—embody oppression, exploitation and discrimination. The Zimbabwe demand "One man, one vote", raised the spectre of majority rule not only in Southern Rhodesia but in the remaining southern footholds of colonialism.

The fact that southern Africa is a coordinated socio-economic complex is not new, and both foreign imperialism and South African (and Portuguese) colonialism have so viewed it.

In Rhodesia, for example, long preparations made by Britain (with US and French political support) for the continuation of white settler rule[1] after the dissolution of the Federation of Rhodesia and Nyasaland were marked by the turning over of the army and the air force to Southern Rhodesia in 1963.[2] But this alone would hardly have sufficed to hold down the Africans, who comprise 95% of the population. South Africa's Minister of Bantu Administration Botha, for example, has indicated that South Africa and Rhodesia had an understanding long before the Unilateral Declaration of Independence (UDI) of November 11, 1965. After the usurpation of power by the Smith régime, the role of Britain and the United States in counselling "restraint" and in holding up the "spectre of violence",[3] permitted the ré-

---

[1] The history is traced back to 1961 by Sir Charles Ponsonby, President of the Royal African Society (and brother-in-law of the former Governor of Rhodesia, Sir Humphrey Gibbs) although he places the main responsibility on the Rhodesian planters, rather than on collusion on the part of the British government. See *African Affairs*, July 1966, London.

[2] The resolution tabled in the Security Council by Ghana, Morocco and the Philippines not to transfer any "powers or attributes of sovereignty" until a representative government was established was defeated by a British veto, with the United States and France abstaining. The *Department of State Bulletin*, October 7, 1963, pp. 559-61.

[3] In retrospect, described as a "myth, in view of the small white minority as against the territory's four million Africans". Keith Irvine,

gime to be reinforced by South Africa, while effective action against it was being hampered on an international plane.

Britain, as the responsible "administering power", advanced a "strategy" of partial economic and financial pressures against Southern Rhodesia alone, rather than the use of force, or even a mandatory embargo with no loopholes. At the Commonwealth Conference in Lagos in January 1966, Prime Minister Wilson blandly declared that economic necessity would bring down the rebel government within "weeks rather than months".[1] Although sceptical, most of the Commonwealth countries were seduced from taking further action.

Washington's "full support" for and assurances of the effectiveness of the British government's limited economic measures—even underwriting colonialist cooperation—were voiced by the Assistant Secretary of State for Africa: "The Portuguese authorities in Mozambique and Angola and the South African authorities have shown a correct attitude. They have respected the British oil embargo and show every sign of continuing to practice their neutrality in what they see as a domestic British problem."[2] But this was so far from the case, that the British representative under world pressure had to apply to the Security Council in April 1966 for authority to employ force against vessels which were carrying oil to Southern Rhodesia—but only via the single port of Beira, Mozambique.[3]

Although the African and Socialist states saw through such diversionary and dilatory tactics and continued to urge the use of force, they were not successful and the US-backed British request[4] was finally approved (10 to 0, with 5 absten-

---

"Southern Africa: The White Fortress", in *Current History,* February 1968.

[1] *The New York Times,* May 6, 1966.

[2] Speech by G. Mennen Williams, January 28, 1966 in the *Department of State Bulletin,* February 21, 1966.

[3] A tanker was prevented from discharging its cargo there, but the bulk of oil shipments was passing through Mozambique by rail from South Africa. The *Department of State Bulletin,* March 6, 1967.

[4] Similarly, Washington backed Britain's semi-official talks with rebel representatives (Ambassador Goldberg: "to investigate any prospect of peaceful resolution". The *Department of State Bulletin,* June 20, 1966, p. 991), which African states attacked as smelling of a "deal" and a big step back from the original London position of no talks until the rebellion itself was quashed. The African resolution, supported by the

tions). Similar compromises, e.g., a ban on exports of chrome and tobacco from Southern Rhodesia, instead of imposing upon her universal, mandatory sanctions (including a ban on oil) and the blocking of transit via South Africa and Mozambique, moved further and further away from African and Soviet proposals for an effective boycott.

The Commonwealth Conference of September 1966 also found the African states resolved that "force was the only sure means of bringing down the illegal régime".[1] But Britain stalled until the end of the year, promising to bring a resolution in the United Nations merely for selective sanctions. The US delegate in the United Nations refused to support African resolutions in October and November because of their "immoderate language and because they impugned Britain's motives".[2] When the selective sanctions resolution was passed in the Security Council on December 16, it did not go far enough nor close loopholes.

The US delegate's backing of Britain "will cost the US considerable in independent Africa,"[3] wrote a perceptive American observer. Selective sanctions do not halt oil and other materials from flowing into Southern Rhodesia, nor keep her from selling products to South Africa for re-export. Africa wanted a much tougher resolution. The Nigerian delegate, described as "one of the most Western-minded of the African delegates," was "as strident as any" anti-West delegation. In African eyes "the U.S. is coupled with the U.K. in a western plot to allow the Smith régime to hang on and eventually to come to terms—its own terms—with Britain".[4] The US Administration is accused "of favoring, for economic reasons, the survival of racism in Rhodesia and South Africa".[5] US vulnerability to charges that it is "a neo-colonial power is greater than it was before the vote". The key role of the US in southern Africa had been pointed out earlier by President Kaunda, for "without particularly ac-

---

USSR urging the British to consult with African leaders rather than the racist minority, failed to pass in the Security Council. The *Department of State Bulletin,* op. cit., pp. 986-91.

[1] The *Department of State Bulletin,* March 6, 1967, p. 372.

[2] Loc. cit., p. 373. At the same time Ambassador Goldberg admitted in his statement of December 12 that Southern Rhodesia represented a threat to the peace and justified mandatory sanctions.

[3] Drew Middleton in *The New York Times,* December 19, 1966.

[4] Ibid.

[5] Ibid.

tive American support",[1] the stoppage of the flow of oil through Mozambique and South Africa cannot be effective.

Within the United States itself there was opposition even to Washington's advocacy in words for economic sanctions, not only from the ultras (Senator Goldwater's successor Senator Paul Fannin of Arizona, Representative James B. Utt in the House, the American South African Committee established in Washington primarily to support the Smith régime), but also from Dean Acheson and the *Washington Post.* A "Peace with Rhodesia" banquet in Washington, attended by important people from American trade and financial circles hailed Smith as "the George Wallace of southern Africa".[2] Sympathy for the African people came from progressive and black Americans, who condemned British and US financial titans for coddling the apartheid that threatens black Africa with a "massive, violent and catastrophic race war".[3]

Events quickly revealed the farce of selective sanctions and, as anticipated, how London and Washington had kept the Smith racist régime viable by buffering its trade and supply routes through the colonial powers of South Africa and Portugal.[4] The moral suasion of the Afro-Asian mem-

[1] *The Economist,* November 19, 1966.

[2] *National Guardian,* November 18, 1967.

[3] A. Phillip Randolph, President of the Brotherhood of Sleeping Car Porters, address to third biennial Congress of the American Negro Leadership Conference on Africa. *Herald Tribune,* January 28, 1967. As contrasted with the criticism of the US position by African leaders, however, he called on American Negroes to support President Johnson's policy of backing UN mandatory sanctions because Britain would not live up to its responsibilities. This ambivalent appeal may not have been unconnected with the State Department's facile change of propaganda emphasis after the Commonwealth Conference in September—of playing down the US role and dissociating as much as possible from Wilson, to escape the brunt of world criticism.

[4] "Insight" in the *Sunday Times* describes part of the mechanism: Shell Middle East was selling crude oil to a refinery in Durban, South Africa (jointly owned by Shell and British Petroleum in London). Refined products were being sold to Shell/B. P. Marketing (South Africa), which sell to dealers who supply Rhodesia. Portugal was re-exporting to Rhodesia under false papers. Rhodesian copper went to South Africa and Europe via a British-based Sales Co. (*Sunday Times,* August 27, and September 3, 1967).

In the first quarter of 1967, Southern Rhodesian export (especially chrome, asbestos and nickel) rose 71% to Switzerland, 50% to the United States, and 38% to Japan. *Newsweek,* November 20, 1967, Atlantic Edition, London. All further references to this source are to the International Edition.

bers at the Commonwealth conferences (including the threat of leaving) and at the United Nations had been of no avail in forcing the hand of Britain or the United States. The independent African states were still bound by strong economic ties to the imperialist powers, who considered them militarily too weak to initiate force of their own.

In the following years, moreover, when they did attempt to coordinate their numerous but as yet unorganized forces[1] against the mutually assisting racist minorities, Britain, for example, sought to tie the hands of neighboring anti-colonial forces by such methods as refusing to sell arms to Zambia and demanding that it not be used as a base for freedom fighters going to the aid of Africans in Rhodesia. This, in turn, encouraged South Africa to make the same demand and even to threaten Zambia with military attack if she refused to comply, i.e., a "preventive" war. At the same time, the hand of Washington was discovered helping the racist régime to purchase weapons and pay Portugal and West German instructors to train Rhodesian military units in South Africa.[2]

If, in the period 1965-70, London with quiet Washington support had been defensively buffering the Salisbury government in the United Nations, on the fifth anniversary of usurped minority rule, the British representative was again to be found applying the veto in the Security Council, but this time against the resolution to keep in force the existing limited sanctions against the racist régime.

Shortly thereafter, Washington made its contribution to this new phase of more aggressive action to bring Rhodesia out of its isolation by granting a licence to Union Carbide, as reported in February 1971, to import 150,000 tons of South Rhodesian chrome ore. Congressional approval of the purchase and importation of such ore followed on November 11, 1971, coinciding to the day with the sixth anniversary of UDI.

This could be considered, perhaps, an intimation that chrome was more than a "raw material" issue or that it was not the cost of imports of Soviet chrome that had caused

---

[1] On August 7, 1967, the African National Congress and the Zimbabwe African People's Union (ZAPU) announced that joint forces of the two organizations had engaged Rhodesian security forces in battle.

[2] Tanzanian newspaper *Ngrumo*, quoted in *Pravda*, January 9, 1968.

Washington, as was claimed, to lift the ban. For, as Deputy Foreign Minister Jacob Malik pointed out in the United Nations, the Soviet Union was selling its chrome at the regular market price and in fact had supplied 49 per cent of US imports even before the UN embargo.[1] Furthermore, an even less broadly advertised fact was that the "US has so much chrome in its strategic stockpile that the General Services Administration proposed that 1.3 million tons—enough for 10 years of defense needs—be declared excess". Thus, if it was Foreign Secretary Sir Alec Douglas-Home who at the end of 1971 had closed the "shameful deal" for recognition of the Southern Rhodesian racists, according to TASS,[2] Washington appeared no less interested than London in their political rehabilitation.

While in Rhodesia, Britain largely out of historical colonial profitable economic ties has remained the predominant foreign patron and US imperialism plays a much lesser role, there is little doubt that in the case of the Portuguese colonies the United States, although not the traditional or primary commercial partner, has gained preeminence over Britain, both politically and militarily since World War II. Economically, since Portugal's initial contact in Angola in 1483, the slave trade was the major source of profit until the mid-nineteenth century; and in Mozambique since 1498—ivory, gold and precious stones, and more recently, the 100,000 contract laborers supplied annually (in accordance with a convention concluded in September 1928 and renewed in 1962) to South Africa, and subjected there to apartheid, for which Portugal receives one-half of the workers' first four-months salary for "transportation". This makes a mockery of the vaunted "assimilation" policy of Portugal, which boasts of equal rights for Blacks in her "overseas territories". Thus, it has been human labor and casual mineral exploitation which not only helped prop up an economically backward and poverty-stricken Portugal (while the biggest mineral wealth remained undiscovered until the 1960's and then was exploited by an influx of foreign capital, including

---

[1] This increased after the ban to 58%. *Herald Tribune,* November 17, 1971.

[2] Statement in *Pravda,* December 4, 1971. It may be noted that there is no possibility provided to the African majority of achieving parity—even theoretically—with the tiny ruling white minority before the year 2035.

the heavy US investments originally in diamonds and extending especially to oil in the past decade), but also provided the socio-economic links with the main reactionary social system below the Zambesi.

Historical responsibility for the roots of the social and racial problems of South Africa, it is true, goes back to the colonial powers' struggle for domination of the African peoples—from the Portuguese discovery of the Cape of Good Hope in the 1480's, followed by the Dutch seizure in 1652, and then the British who ousted the Dutch from the Cape during the Napoleonic Wars and sent out settlers in 1820. The Dutch-descended Afrikaners, who moved inland, established two Boer republics in the 1850's, which British colonialism subdued in the Boer War of 1899-1902. But not before Cecil Rhodes' British South Africa Co. had taken over Rhodesia and suppressed Africa's Matabela and Mashona uprisings against white settler rule.

In more modern times, from the dual British and Boer inflow of settlers, the Union of South Africa was established in 1910, with the Afrikaner emerging predominant after World War II. Colored and Indian people of South Africa often looked to the British for assistance against the virulent racialism of the Afrikaner. Although contradictions existed between the two, as the Communist Party's Program "The Road to South African Freedom" (adopted in 1962) pointed out: "In the oppression, dispossession and exploitation of the non-Whites, British imperialism and Afrikaner nationalism found common ground."

Today, moreover, to these immediate active "partners in apartheid" must be added a third, relatively silent partner (see Chapter "US Neocolonialism", section "Economic Basis") —the United States, especially since 1960[1]. This stems fundamentally from American monopoly ties to the socio-economic system and the capitalist class of South Africa, which dominates and exploits some 18 million non-whites—Africans, Colored of mixed origin, and Asians in southern Africa through the system of oppression called apartheid, built on the continent's biggest white base of privilege, nationalism and chauvinism.

---

[1] Since 1962 the UN General Assembly, which previously had regularly merely condemned apartheid, urged its members to institute a diplomatic and economic boycott to force a radical change in South African policy. Resolution 1761 (XVII), November 6, 1962. By 1966

Table VI

Southern Africa, population (rounded)

	Population (in mill.)[1]	% *Whites*[2]
South Africa	19.6	19.3
Mozambique	7.4	1.1
Angola	5.4	3.5
Southern Rhodesia (Zimbabwe)	5.1	4.6
Lesotho	0.9	0.3
Botswana	0.6	0.06
Namibia	0.6	13.9
Swaziland	0.4	2.5

Source:

[1] *Demographic Yearbook 1970*, 22d issue, Population Trends, UN, N.Y., 1971.

[2] *Demographic Yearbook 1963*, 15th issue, Population Census Statistics II, UN, N.Y., 1964.

The virus of racism has been spread and has infected even the working class, weakening organized white labor which has succumbed to segregated unions.[1] The Africans have no political rights—neither to vote, nor assemble, nor form trade unions; must carry passes from their segregated slums (sooner concentration camps) to their place of work as cheap manual labor in a white town; are socially treated and "taught their place" as inferiors, in effect sub-men; are subject to beatings, arrests without an appearance before a judge for 180 days, and torture.

Social oppression, by its nature, has far-reaching politi-

the Assembly asserted that "universally applied mandatory economic sanctions" were the only road to a peaceful solution, and that South Africa had been encouraged to pursue its disastrous policies through the persistent opposition to such sanctions by its "main trading partners". Resolution 2202(XXI), December 16, 1966.

[1] See A. Zanzola, member of the Central Committee of the Communist Party of South Africa, "The Conscience of South Africa" in *Pravda,* July 29, 1966; also R. E. Braverman, "Trade Union Apartheid" in the *African Communist,* No. 29, 1967. In the first quarter of the century, when opposition to British imperialism and mining capitalism came largely from Afrikaner nationalist workers in the Rand, class and color posed some difficult problems for the Communist Party, too. A later reassessment by the Communist Party put greater emphasis on racial oppression as the major criterion for the struggle against capitalism and for the setting up of a non-racial Socialist state. On the other hand, the non-white élite apparently hoped for a deal with British imperialism and Afrikaner capitalism.

cal implications. Racism and chauvinism flow over and are directed against other ethnic and social groups as well, including whites, e.g., anti-Semitism is widespread and the English are deprecatingly referred to as Anglo-Jews. Liberals and Communists are persecuted, democratic and Socialist literature (including Marxist classics), with their emphasis on racial equality, are prohibited. This is not surprising for leading South African officials have had close ties with fascism and its ideology.[1]

The US ultra-Right, in turn, quite naturally draws encouragement from and identifies itself with the rulers of South Africa. With the appearance in 1965 of "Apartheid and U.N. Collective Measures: An Analysis" published by the Carnegie Endowment for International Peace—a volume weighing the feasibility of sanctions—American ultras labelled it a "battle plan for U.N. invasion of South Africa".[2] Rushing into the ideological fray were the *Chicago Tribune*, Goldwater, a newly organized American-African Affairs Association with Representative J. Ashbrook (Republican, Ohio), Ralph de Toledano and Max Yergan, who toured Rhodesia. The AAAA's public relations man was Marvin Liebman, who also helped organize the Committee of One Million against the admission of the People's Republic of China to the United Nations, the American Committee for Aid to Katanga Freedom Fighters, and the supporters of Goldwater's nomination for the presidency. The *National Review* also took part in the campaign and 15 of its editors and contributors were listed among the 54 names on the prospectus of the AAAA. For their part, the "conservatives", as the ultras of South Africa's ruling Nationalist Party call themselves, had given their vocal support in the 1964 US presidential elections to Goldwater, whom they lauded as an "exponent of an international creed".[3]

[1] The chief of police, Van der Bergh, was a well-known Nazi. M.C. Botha, Minister of Bantu Administration, and Albert Hertzog, Minister of Health, are in the fascist wing of the Nationalist Party. Piet Botha, Minister of Defence, is a Nazi sympathizer, along with Prime Minister Vorster and Minister of Finance Deidrich (who had close ties with Herman Abs, key man in Nazi and postwar West German banking, industry and armaments). *Comment,* February 10, 1968.

[2] Vernon McKay, "Africa and the American Right" in the *New Republic,* March 26, 1966.

[3] S. Uys, "Goldwater in South Africa" in *New Republic,* January 6, 1968.

Washington's attitude toward social oppression has been more circumspect. US official policy in the early sixties, taking into account the upswing in the African national-liberation and world progressive movement as a whole, was forced to show more ambivalence than previously toward South Africa's racism.[1] Washington could be expected to join regularly in what it considered a "ceremonial condemnation of apartheid"[2] (admonitions of South Africa, declarations that its racial policy was "repugnant" etc.), but also to water down African proposals for effective action (e.g., describing the situation as "seriously disturbing" instead of "threatening" international peace and security. The latter would have given the Security Council decisions mandatory rather than recommendatory force).[3]

The contradictions of the "New Frontiers" policy between its imperialist substance and verbal tightrope walking was mirrored in Nairobi, when G. Mennen Williams, Assistant Secretary of State for African Affairs, declared on a visit in 1961 that the United States stood for "Africa for the Africans". This so shocked London and the white racists that it led him to "explain" that he had not meant only for the black Africans. And, when in Uganda soon thereafter, Mr. Williams warned Africans not to exchange one tyranny for another, he felt impelled to issue immediate assurances that he had not meant to characterize British rule as tyranny.[4]

A hardening of foreign policy following the assassination of President Kennedy represented more than a change of occupants in the White House. Consideration for American monopoly's growing economic stake in southern Africa paralleling such successes as the US-promoted political and military moves in the Congo, was reflected in a decreased sensitivity to the African struggle against racialism.

---

[1] Following the Sharpeville police massacre of 72 Africans peacefully protesting against segregation "pass laws" on March 21, 1960, world opinion was so aroused that even the State Department had to issue a statement regretting the tragic loss of life. The *Department of State Bulletin,* April 11, 1960, p. 551.

[2] Arthur M. Schlesinger Jr., *A Thousand Days,* Boston, 1965, p. 580.

[3] *United Nations Review,* August-September 1963, pp. 20-24.

[4] In commenting on this dilemma, the outstanding Africanist Professor Emerson of Harvard noted that a Communist "would have said 'Africa for the Africans' and 'tyranny' and stuck by it". "American Policy in Africa" in *Foreign Affairs,* January 1962.

In 1964, for example, President Johnson sent a close friend, Charles Engelhard, who perhaps more than anyone else symbolized the US-South Africa gold bond with apartheid, as his representative to the Zambia independence ceremonies.

Against the growing demand for sanctions, the State Department polemized that a legal basis was lacking, and in any case, sanctions would be ineffective; moreover, the United States was not going to drive South Africa into "isolation" by breaking off commercial relations, but was seeking "peaceful accommodation of the forces for change". To the charge that US growing economic ties were helping to sustain apartheid, Assistant Secretary of State Williams evasively replied that US aims in South Africa were "essentially political".[1] This did not refute the charge nor meliorate the US position.

President Johnson's move Rightward with respect to South Africa's system of apartheid was frequently countered in Establishment writing by an alleged US "liberal" official position in the South West Africa case. But this is quite dubious even though the US role is not easily unraveled. With the United States and other Western powers advising Africa to take no action, nor to deviate from strict legal procedures,[2] Ethiopia and Liberia had been trying since 1960 to confirm earlier advisory opinions of the International Court at the Hague that South Africa was violating its League of Nations mandate by the extension of apartheid into that territory. Although the World Court had ruled in 1962 that Ethiopia and Liberia (the only two black African states in the League of Nations, which granted the mandate in 1920) had "standing" (i.e., sufficient legal interest in the subject to bring their complaint), a differently composed Court[3] decided on July 18, 1966, that these two countries

[1] Assistant Secretary of State Williams in the *Department of State Bulletin,* March 21, 1966, pp. 432-39; and Secretary of State Rusk in *The New York Times,* July 22, 1966.

[2] See R.N. Nordau, "The South West Africa Case" in *World Today,* March 1966. The fundamental weakness of seeking to disentangle a colonial legal web through procedures established by the Western powers themselves had been astutely noted and rejected by Prime Minister Nehru when India took action to incorporate Goa, Damao and Diu on December 18, 1961: "Colonialism is illegal."

[3] For details of political and legal maneuvers see R. First, "South-West Africa" in the *Labour Monthly,* September 1966; B. Pilkington in the *National Guardian,* July 30, 1966.

lacked "standing". The Court thereby declined on a technicality (the point on which disqualification of the two parties was based was not even advanced by South Africa's lawyers in their final submissions) to rule whether the mandate remained in force, and whether the extension of apartheid, as *The New York Times* ironically put it, "promotes material and moral well-being of 526,000 inhabitants (of whom 75,000 are white)".[1]

Although the US representative on the Court had managed to vote with the dissenting minority without thereby changing the outcome, it was more than suspicious that the State Department "found virtue in this anticlimactic end to years of deliberation", and that a statement by the US Mission to the United Nations "seemed to heave a sigh of relief that a confrontation with Cape Town had been avoided".[2]

That the Court decision would not have altered matters but merely helped Britain and the United States continue their interest-based delaying tactics[3] was anticipated on the eve of the ruling by a revelation of US plans: if South Africa's claim to administer the territory is upheld, according to *The New York Times*, US policy will continue to plead with South Africa not to spread apartheid to South West Africa. If, on the other hand, "the Court recognizes U.N. responsibility for the region, Washington will then press the other African nations to bide their time until South Africa has had a chance to comply with new international standards".[4] Most African nations "backed in the U.N. by Asian and Communist countries would soon demand forcible action against South Africa, which London and Washington have thus far opposed".[5]

---

[1] *The New York Times*, July 21, 1966.

[2] *Nation* (editorial), August 8, 1966.

[3] Economically, the territory's diamond and copper resources (being exploited by South Africa, the United States and Britain) were estimated to be able to last for less than a generation. Direct economic interests are only a minor part of the entire southern complex. Strategically, according to Richard Gott (in the paper "South West Africa: The Defense Position" read to the International Conference on South West Africa held at Oxford in March 1966), the territory encircles Botswana, provides a link to Rhodesia and such a forward base as the Caprivi air strip some 500 miles closer to independent Africa than the border of the northernmost province of South Africa. *Labour Monthly*, loc. cit.

[4] *The New York Times*, July 18, 1966.

[5] Ibid.

And this, indeed, they continued to do in the years following the General Assembly declaration that the mandate was terminated and that "South West Africa comes under the direct responsibility of the U.N.".[1] By 1969, moreover, in contravention of this resolution, Prime Minister Wilson was reported to have approved the secret purchase of uranium ore by a Rio-Tinto Zinc Corporation subsidiary from South Africa and Namibia.[2]

Washington's more guarded position consisted in ineffectual declarations that South Africa's "illegal occupation" of Namibia could not be condoned and in mildly discouraging investment and trade by not granting US government guarantees.[3] The United States felt compelled to go along with the Security Council resolution[4] which recognized the General Assembly's decision to terminate the mandate and declared South Africa's authority in Namibia illegal. The resolution's reaffirmation of the arms embargo against South Africa, however, was still far from the political and economic coercive measures which Africa and the Socialist states felt were needed.

In less publicized ways, Washington has been making available to South Africa access to research and expertise, which has not only a direct bearing on economic and military strength, but also indirectly upon apartheid. Thus, in 1961, when the US National Aeronautics and Space Administration (NASA) established a Radio Space Research Station, which included the Jet Propulsion Laboratory operated under contract with the California Institute of Technology, the State Department gave assurances that the United States would not be a party to apartheid arrangements. Circumvention took place by contracting the operation of the $2.5 million station to the South African quasi-governmental Council for Scientific and Industrial Research (CSIR), rather than by directly employing Americans and

---

[1] Resolution 2145 (XXI), October 27, 1966. As against the Afro-Asian proposal to place the territory under the United Nations, or the Soviet proposal to declare it independent, the US position was to set up a commission to recommend a "timetable".

[2] *Sunday Times*, July 1970.

[3] Ambassador Charles Yost at the UN 25th anniversary celebration, May 20, 1970.

[4] Senate Resolution 287 adopted on July 29, the *Department of State Bulletin*, September 7, 1970. Similarly, the International Court's advisory opinion on June 21, 1971.

local nationals as are most NASA stations abroad.[1] The approximately 250 CSIR employees, trained in the United States, operate under apartheid regulations, which specify that non-whites do not hold jobs above the menial level, nor receive equivalent pay for the same work, nor dine or share public transportation with whites. NASA funds also partially finance a four-year training course in electronics for whites established in 1964, according to the deputy director of the station, who was formerly commanding officer of the South African Air Force School of electronics.[2] Participation in space research has brought national prestige to racist South Africa, membership in the international Committee on Space Research, and the development of skills in the latest equipment and electronic techniques.

The United States is linked with the economic and derivative social aspects of metallurgical problems connected with South Africa's most important industry—gold (in which the United States is the biggest gainer from the artificial low price of gold). Only the richest veins are worked, and these to record depths of over 12,000 feet. With the price inflexible, greater profitability is gained through increased exploitation—essentially by lowering production costs both in technology and in human labor. In the former sphere, significant cooperation takes place with the University of Minnesota (e.g., the director of the Mining Research Laboratory of South Africa's Chamber of Mines is also a professor at the University of Minnesota), Massachusetts Institute of Technology, and the University of California at Berkeley. In the search to reduce further the cost to mine owners of the nutritional requirements of the annual turnover of some 222,000 African laborers who work in the mines (which smacks of a study of food requirements of slave labor), South African researchers maintain close relations with American organizations, including Massachusetts Institute of Technology and the US Department of Agriculture.

In the field of atomic energy, an examination of the close scientific and technical links with the United States shows scores of scientists to be American trained, especially at the Oak Ridge National Laboratory. The single reactor at Pelindaba is an Oak Ridge design purchased through Allis-Chalmers.

[1] D.S. Greenberg, "South Africa" in *Science,* July 10, 1970.
[2] Ibid.

4. IDEOLOGICAL FORCES

Although dialectical materialism does not attribute a direct linear relationship between socio-economic formations and their superstructure, including ideology, nevertheless, since Marx and Engels hammered out the importance of the economic basis in historical development there is a tendency for some to conceive of this basis as a precondition, if not predeterminate, of all social development. By implication at least, there is a denigration of the force of ideas and ideology.

Lenin, in seeking to combat a static, compartmentalized or schematic approach to change in the underdeveloped countries pointed out, for example, that for them the capitalist stage is not inevitable if one takes into account the force of ideas ("systematic propaganda") and the "aid of the proletariat of advanced countries".[1] Furthermore, the ideas themselves develop in the process of struggle and under the influence of world forces and events. In this connection he foresaw that the "movement of the majority of the world's population, originally aimed at national liberation, will turn against capitalism and imperialism and will, perhaps, play a much more revolutionary role than we have been led to expect".[2]

This, in effect, has since been demonstrated by a number of African countries now taking the non-capitalist path. But, the ideological struggle, like that in the other spheres, is a two-way proposition, of which imperialism is quite aware. In this connection, the ideological efforts of the United States in Africa are also not to be underestimated as a reactionary force.

Ideology, to be sure, has aspects which are both objective, i.e., reflective of actual conditions, trends, policies and actions, as well as subjective, divorced from or at variance with the latter or simply evolved at a writing desk "to influence the minds and hearts of men". It is not difficult to understand why the latter particularly would appeal to Washing-

---

[1] "Report of the Commission on the National and Colonial Questions to the Second Congress of the Communist International, July 26, 1920" in V. I. Lenin, *The National-Liberation Movement in the East,* M., 1957, pp. 268-69.

[2] "Report to the Third Congress of the Communist International, June 5, 1921", op. cit., p. 290.

ton and US monopolies as a cheap answer to knotty political, economic and social problems.

Historically, it is a commonplace for reaction in America, e.g., ultra-Right organizations such as Sons and Daughters of the American Revolution, to seek to identify itself verbally—now that it is safe—with the progressive events of the country's past. Similarly, in the case of Africa, frequent reference has been made by US officials to the anti-colonial and democratic traditions of America, and to the links between the two continents—even extending to the "bond" with Africa, as the source of American slaves. No distinction, to be sure, is made between the bonds of the US ruling classes and former slaveowners with Africa, as contrasted with the bondage of the Afro-Americans.[1]

US officials place great emphasis, too, on the compound lie of anti-Communism—first the distortion and falsification of its content to make it appear reprehensible, and then the labelling of all patriotic forces as Communists; on the industrial strength of the US economy to overawe, lure, or by association to make the capitalist social system appear in a better light; and on efforts to absolve class society in the United States of the onus of racial oppression, exploitation and discrimination by ascribing such shortcomings to mankind in general, rather than to capitalism, as organic features.

Many of Washington's propaganda lines have been taken over from the generally liberal policy recommendations (which, as policy, have generally not been adopted) made by educators, such as those of Northwestern University's Program of African Studies,[2] e.g., to speak in terms of African interests, to recognize an African policy of non-alignment in the cold war, to support self-rule and racial equality rather than colonial rule, to encourage development free from outside interference.

Although variations of these concepts are to be met in Washington's "war of words", raising the bogey of Communism (once favored by the Third Reich) remains a pivotal

---

[1] Compare a more blatant British colonial apology: The slave trade "was based on the sale of slaves by other Africans to white slavers in return for goods which those Africans wanted ... the insistent African hunger for the manufactures of the more advanced civilizations." Scipio, *Emergent Africa*, pp. 41-42.

[2] Published as *United States Foreign Policy*, 1959, op. cit., pp. 13-17.

part of US cold war propaganda. Liberal government advisers and critics, when dealing seriously with US official anti-Communism, have asserted, for example, that in Subsaharan Africa Communist influence is minimal and that the cry of Communism is, indeed, "most often heard in the African countries whose commitment to the West is strongest—South Africa, the Federation of Rhodesia and Nyasaland and the Portuguese territories.

"This, however, scarcely makes Communists of African nationalists. When we are told that Dr. Hastings Banda, the leader of the Nyasaland African Congress, was influenced by Communists because he met certain Russians at Accra — the Russian delegation there totalling 6, as against more than 100 Americans—we may well ask for specific proofs of this influence."[1]

Whatever the merits of such liberal criticism, however, it also implies certain falsehoods: that the ideas of scientific Socialism and the Socialist path are not universal but Russian inventions; are not matters of choice for the African people and states themselves, but subject to Washington's decision; and are not influential, hence do not justify US counter-action.

In fact, the US imperialist big stick, in one form or another, has been masked behind anti-Communism—from North Africa and the Middle East to the Congo. When this transparent line, however, began to wear thin after the US (airlift)-Belgian (paratroops) and British (airbase) intervention ("rescue operation") in the Congo in November 1964, the verbiage changed from the need to "contain Communism" and "fulfill our commitments" to a more abstruse justification of US intervention "whenever its absence will create regional instability of expanding proportions", as formulated by an influential member of the State Department's Policy Planning Council.[2] Intervention, he opined, "has to be judged largely on its international merits not in terms of specific domestic consequences within individual states".[3]

[1] Op. cit., p. 11. In the decade since then, it may be wondered how many such charges have been deliberately leveled as part of psychological warfare to confuse, intimidate and divert.

[2] Zbigniew Brzezinski, "The Implications of Change for U.S. Foreign Policy" in the *Department of State Bulletin,* July 3, 1967.

[3] Apparently, a reformulation of the obsolescent "domino effect". Thus, he noted, "it is that distinction which warrants our involvement in the effort to create regional stability in Southeast Asia". (Ibid.)

Thus, another formula to rationalize the US postwar policy of policing the world.

Nonetheless, behind all of these catch-all phrases, US imperialism has been very much concerned with African leaders and ideologies looking to a non-capitalist path to Socialism.[1] This is especially relevant to scientific Socialism, looking forward to a working class base for economic development under modern conditions, free of imperialist influence. In contrast, "African Socialism",[2] harking back to a communal and egalitarian traditional society, holds that Africa unlike Europe is free of rival classes, has a unique culture and history, and therefore must depend entirely on Africa's building its own future. It is not difficult to understand that such romanticized backward-looking views, which also would cut Africa off from progressive world forces and allies, are not regarded as a real threat by imperialism and have even been used by reaction in opposition to scientific Socialism.[3]

US domestic industrial and economic might has provided Washington's "idea men" with a strong argument for promoting American capitalism abroad. Not that US foreign investments and trade have brought Africa anything comparable with what has been drained out by American monopolies. It is this fact which programs of private capital[4] and US government such as "aid" seek to cover up. Part of

---

[1] In tropical Africa, for example, an early advocate, Nkrumah, in his book *Consciencism* (1964), held that ideology is central to African revolutionary struggle.

[2] With various shades of emphasis: President Senghor's related spiritual concept of "Negritude"—"...Negro-African society is collectivist or, more exactly, communal, because it is rather a communion of souls than an aggregate of individuals". (*African Socialism,* London, 1964, p. 49.). President Nyerere emphasizes the fact that nobody starved in traditional society because he could depend on the community as constituting the essence of Socialism. (*Ujumaa: The Basis of African Socialism,* 1962, p. 3.) See I. Cox, *Socialist Ideas in Africa,* London, 1966.

[3] "Negritude", according to Nkrumah, "serves as a bridge between the African foreign-dominated middle class and the French cultural establishment". *Class Struggle in Africa,* International Publishers, N.Y., 1970.

[4] Numerous avenues pursued by US monopolies in "sowing goodwill" range from the activities of their Foundations to direct "community relations". In mid-1970, e.g., Mobil Oil Co.'s subsidiary in Ghana sponsored a nationwide painting and sculpture contest of the arts and education, traffic safety programs, and local employee involvement in community tasks.

the annual psychological game revolving about appropriations is for Congressmen and the communications media, which are essentially organs of big business, to charge that aid amounts to "giveaways" and "wasteful expenditures on prestige projects". Then, the administration generally seeks "to woo" Congress and US monopolies by its stress on US "national security" and economic benefits derived; and, to a lesser degree, on moral and welfare principles—in order to make an impact on African and world opinion. The aid itself, whose primary significance lies in its political strings and economic motives (see section "Economic Basis"), does involve minor concessions but hardly enough to evade past and present imperialist responsibility for the continent's economic backwardness.

If the US possesses economic strength achieved on the basis of technology and exploitation (in part in Africa), capitalism has generated weakness in the social sphere, where US imperialist-African contradictions are glaring, e.g., racism in southern Africa (and the comparable domestic discrimination against African-Americans). In the main, US officials have sought to absolve imperialism of the onus of support for racism by separating out—as if unrelated—the question of the profitable monopoly ties of the United States from its attitude to world action against apartheid. Thus, the US African Policy Statement declares that the United States does not believe in cutting ties "with this rich, troubled land" and will "continue to make known to them and the world our strong views on apartheid".[1] Another tack has been to argue that sanctions, boycotts, etc. are ineffective.[2] It is well known, however, that Washington has not at all hesitated to employ economic weapons against states with which it differs ideologically. The US line in the United Nations has been to raise the spectre of violence and bloodshed in southern Africa and therefore to urge "modera-

[1] *US and Africa in the Seventies,* op. cit., p. 521.

[2] Prime Minister Heath, in a talk with President Kaunda of Zambia, as reported from London, even more hypocritically "trotted out the tired lie about economic progress in South Africa eventually compelling the abandonment of apartheid (South Africa has had a continuing boom for over 20 years, and throughout that time apartheid and its attendant evils have gotten worse)". And he indicated, as if a finishing stroke, the non-sequitur that "Zambia receives aid from China and Russia". *Herald Tribune*, October 20, 1970.

tion" and gradualism,[1] i.e., to block world action, while simultaneously expanding economic relations. Furthermore, as if in justification for not bringing pressure to bear on South Africa, US apologists have admitted, disingenuously, that the record on American domestic racial relations also is "not unspotted". A cold war decoy which has been employed to divert attention from the issues involved is to claim that the Russians, Chinese or Cubans are "exploiting" the situation.

Differences in emphasis and tone are to be noted in US apologetics since MacMillan's "winds of change are blowing" speech made at a high tide of African national liberation in 1960. Thus, after South Africa had quit the Commonwealth in 1961 on the question of apartheid, a US delegate at the United Nations warned South Africa that its racial policies could "rock the entire Continent".[2] It seemed credible then that Washington, as claimed, was seeking through diplomatic approaches to Praetoria to bring about some change more in keeping with the times.

By the second half of the decade, however, the US official line moved to the Right, more often stressing the rationale of US accommodation to the status quo, echoed duly even by press organs considered liberal on racial issues: Although South Africa's race policy is abhorrent, it is "her own affair —like Spain's where US has military bases, and Portugal, a US ally."[3] Prime Minister Vorster says the British have not solved their race problem and American racial policy "is no concern of ours". The military embargo is a "pinprick", they get arms elsewhere. Not giving shore leave because of apartheid to the crew of the US aircraft carrier Roosevelt in February 1967 gummed up a "friendship visit". Encouraging Africans against apartheid makes "Afrikaners here simply crawl together", according to Mrs. Helen Guzman, the "sole parliamentary voice of real opposition"[4] in South Africa.

[1] With respect to the Portuguese Territories, for example, the US African Policy Statement declares that the peoples have a right to self-determination, US will encourage "peaceful progress to that goal", and that the declared policy of Portugal for racial toleration "holds genuine hope for the future", (ibid.). (This, when the African people are already fighting for this right, having been denied it for four centuries!)

[2] *The New York Times*, October 25, 1961.

[3] C.L. Sulzberger in *The New York Times*, December 14, 1967.

[4] Ibid.

The US policy of curbing international action frequently has been further rationalized along pragmatic lines. UN Ambassador C.W. Yost, in a speech to the United Church of Christ in Boston, criticized as "unrealistic" efforts to impose sanctions on countries in southern Africa, even though the denial of human rights was "reprehensible".[1] Artfully, he argued, the failure to achieve results in Rhodesia should be a warning against weakening the United Nations by attempting to apply sanctions against stronger countries such as South Africa.

In Canada, which has pursued a policy similar to that of the US, a stronger domestic opposition forced a study of the government's relations to southern Africa and the issuance after two years of a White Paper on foreign policy. Prime Minister Trudeau commented in March 1970: "I'm not overly proud of this policy. It's not consistent.... We should either stop trading or stop condemning."[2] But there appeared little government enthusiasm for infringing upon monopoly profits in favor of action on moral principles.[3]

In view of growing world censure and greater recognition of racialism in the United States and in southern Africa as a continuing feature of imperialism and class society, which Socialist societies have eliminated, US propaganda media undertook a deliberate campaign in the sixties to spread the myth that Soviet republics and nationalities also were the objects of discrimination and that race problems could not be solved overnight; or even the more sophisticated line that

---

[1] *Herald Tribune*, July 1, 1969.

[2] *Africa Report*, October 1970. The same disingenuous argument is used by the former Under Secretary of State George W. Ball: "Either we should hermetically seal South Africa off from the world," or seek "...to bombard South African society with the free force and play of humane ideas". He then points out approvingly that Houphouet-Boigny of the Ivory Coast is now urging the black states to undertake direct talks with South Africa and there is a growing realization that the presence of 30 black African embassies in Pretoria might encourage the growth of more relaxed social attitudes." *(Newsweek,* November 16, 1970).

[3] Opposition to government policy comes from a group of churchmen, voluntary organizations, trade unionists and returned Canadian University Service Overseas volunteers, who prepared a Black Paper offering an alternative Canadian policy in southern Africa—an end to Commonwealth preferences to South Africa (a decade after her leaving), the discouraging of Canadian involvement especially in the Cabora Bassa Dam and in Namibia, the stopping of NATO aid to Portugal, and the offering of aid to Tanzania, Zambia and Botswana.

an attribute of a bourgeois-democratic society was that it did not attempt to impose equalitarian practices upon all of its citizens but hoped that with time racial discrimination would disappear. The argument, for all its seductiveness, suffered from the flaw that under capitalism there exists an entire framework (political, economic, social and legal) which permits—not to say encourages— the preaching and practice of racial discrimination, which under Socialism has been abolished.

* * *

Recognizing the importance of ideas in influencing the course of events, US imperialism has invested no small amounts of money and effort in distinctly ideological programs, comprising such related fields as information and propaganda, education, religious and cultural activities, technical assistance. In all these spheres, quite evident in information—and not at all labelled propaganda, there is a mixture of fact and fiction, truth and lies, reality and myth.

Foreign communications media, in this respect, are especially significant in Africa, where recently[1] seven countries still had no daily press except government handouts, 15 had no daily newspapers. In all of Africa, there were some 200 newspapers, fewer than 40 national radio systems (plus 10 stations serving racist or colonial regimes). There were about 6 million radio sets (over one-half in Arab-speaking countries), and 20 television stations,[2] with about 600,000 African licence holders (of which 375,000 were in the UAR).

US information is channeled through a vast network of overlapping governmental, quasi-official and private undertakings, quite on a level with American primacy in foreign investment and commerce. Although US news services are mainly in private monopoly hands, 70% of all radio stations in the United States are under federal control, and 40% exclusively for governmental agencies.[3] The Defense De-

[1] Ralph Friberg, "Some Aspects of News from Africa" in the *Democratic Journalist,* No. 6, 1969.

[2] Broadcasting programs follow in the footsteps of the technology of installation, e.g., the first television system in the Congo (K), installed by RCA, operates on the US 625-line standard, rather than the French 819-line.

[3] See Herbert I. Shiller, *Mass Communications and American Empire,* N.Y., 1970.

partment alone has overseas some 38 television and 200 powerful radio broadcasting stations. Voice of America broadcasts from Addis Ababa, Monrovia, Kinshasa, Tangier, as well as from the United States (Greenville, South Carolina).

Needless to say, there is a close correlation in the governmental and private enterprise (military-industrial complex) propaganda line. The US Information Agency (USIA), with its over $100 million a year budget (about 10% of which is designated for Africa), has issued clear instructions that it is not in the business of furnishing news, literature or entertainment for its own sake but the "skilful and inventive presentation of facts" to promote US policy and aims. Former USIA Director Frank Shakespeare not long ago reiterated[1] that he was an advocate of a "hard line" against Communism. USIA publishes its own *American Outlook* in Accra and Kinshasa, and at the same time distributes free copies of *Newsweek*, *Life* and *Ebony*. It has its own teletype wire service since 1960 (linking distant countries, such as Ethiopia and the Ivory Coast)—supplied free of charge to African newspapers. It has its own reading rooms (of up to 5,000 volumes, plus press) in over 30 cities; and furnishes motion pictures (documentaries are prepared by "Today" in Addis Ababa), which have nevertheless met opposition from several independent-minded countries.[2]

In place of Lagos, which for a long time was a major USIA advance post of anti-Communism, Accra began coming to the fore after the February 1966 coup d'état. In October 1966, for example, there was established a Council to Combat Communism in Africa (whose head declared he was unafraid of the label of "agent of western imperialism").

The United States shift out of Lagos is an instructive example of Washington's political line and its ideological ramifications. Nigerian hostility to American imperialism intensified during the three-year war, when US newspapers, radio and television demonstrated an obvious bias in favor of Biafra. (Furthermore, Washington's line continued after the war ended on January 15, 1970, when the US Congress

---

[1] ABC television broadcast, December 20, 1970.

[2] Thus, Egypt and Tanzania forbid anti-Soviet films, and Somalis, for example, have demonstrated against their pro-Vietnamese war content. *International Affairs*, No. 9, 1967.

in March insultingly rejected a request to train a number of Nigerian soldiers in technical subjects—the Federal Government had built an army from 10,000 to some 300,000.) Clearly the position taken by US mass media during the war had not been accidental. In retrospect, Lt. General Gobon, head of state, commented that most of Europe and the United States wanted to break away Biafra from Nigeria since dismemberment suited their economic interests.[1] Therefore, according to Gobon, Nigeria's position did not receive fair treatment in the American press during the war, and Nigeria could not even present its point of view in a paid advertisement in the London *Times*.[2]

Although Washington suffered a political defeat in its relations with Lagos as a result of its official and unofficial activities in Nigeria, the United States has maintained a continuity in its unofficial ideological channels of exerting influence through such organizations as the Foundations, which are not hampered by public scrutiny, can act fast and also on a long-term basis. Thus, for example, Wayne Fredericks, former Deputy Assistant Secretary of State for African Affairs (1961-67), is now head of the Middle East and Africa Division of the Ford Foundation (from which he had originally come to the US government). The latter has provided $120,000 to the International African Institute in London for coordinating research and information. In related educational and technical fields, it has provided financial aid and personnel: several consultants to the government, more than $100,000 to the Nigerian Institute of International Affairs, Ibadan, and about $10 million to the Institute of Tropical Agriculture (another contributor is the Rockefeller Foundation).[3] A Library of World Research, sponsored by Ford and Rockefeller Foundations, together with US and Canadian aid agencies, was to be completed in 1972. Among the program studies, it may be noted, population control figures prominently.

In the field of educational, language and cultural activities, ideological influence is on the whole more subtle and indirect than in the information sphere, and during the pre-independence period centered in religious hands. The foun-

---

[1] *Washington Evening Star*, August 28, 1970.
[2] Ibid.
[3] *West Africa*, June 13, 1970.

dations of the present African school system were laid by nineteenth century missionaries and bear the pattern of academic education, the preparation of clerks to serve in colonial bureaucracies, and a minute university-trained élite oriented to the metropole in the British and French colonies (the more utilitarian Belgian emphasis was on primary education). Although education has been largely secularized since independence, because of the cost, only Guinea had entirely nationalized the school system by the mid-sixties.[1]

Since most present-day African intellectuals and leaders have been educated in Western missionary schools, the influence of the Church is something to be reckoned with. An estimated 37% of Africans belong to traditional religions, 40% are of Islamic faith (about one-half in northern Africa), and 22% Christians (of which, 9% are Catholics—about 22 million; and 7% Protestants—about 15 million).[2] Catholics are mainly in the former French colonies, and Protestants in Liberia, Ghana, Malawi, Rhodesia, South Africa, where the British-US influence predominates.

US church organizations, today, support several thousand missionaries, making them the largest group of American civilian residents in Africa. The scope of missionary influence beginning with education extends to government, the professions, industry and even agriculture. Nevertheless, the wide discrepancy between the ideals preached by religion and the racial discrimination practised in its name or framework has led to secession from the mission churches. In South Africa, for example, there are about 3,000 separatist churches.

Although independent Africa considers capital formation one of the most important single factors in economic growth, the development of skills through education and training as well as for its ideological content has a high priority. Consequently, the new governments early embarked on great expansions and secularization of their educational programs and Africanization of curricula aimed to overcome illiteracy rates of 80-90%, to transform the traditional peasant into a

---

[1] *Education and Nation-Building in Africa,* ed. by L. G. Cowan, London, 1965, p. 16.

[2] Figures compiled on the basis of the *U.N. Demographic Year Book, Europa Year Book, Britannica Book of the Year* and *Whitaker's Almanach,* see also Г. А. Шпажников, *Религии стран Африки (Religion in African Countries),* Moscow, 1967, pp. 38-39.

modern farmer, and to overcome the shortages of trained personnel.[1]

The United States, which lagged behind the European powers in exerting influence in these fields, made rapid strides to catch up. Whereas at the close of World War II American universities conducted programs for all the major areas of the world except Africa, by the 1950's the Carnegie and Ford Foundations had helped to launch graduate African programs at a number of major universities. By the mid-sixties, there were some 42 African programs in American universities, 30 US official textbooks and scientific centers in Africa, which distributed about 600,000 textbooks in 1964; and 30 American foundations were active.

The training of African students in the United States, which began in the late 1950's, involved 2,800 in 1960-61, and jumped to 6,800 (including 1,280 from the UAR) by 1964. In the second half of the decade, these students constituted about 8% of the foreign students in the United States. About two-thirds of the African students are sponsored by the US government, their own government and private agencies (church or foundation). An increasing number of students ("Participants") are pledged to take part on their return to Africa in projects in which US private industry and government are interested. These were to number 4,700 in fiscal year 1966, an increase of 1,300 over 1964, of which about one-half was in agriculture and one-fifth in education. US educational programs have not overlooked potential opposition leaders even in countries with the governments of which the United States has very friendly relations, e.g., a scholarship program for students (mostly refugees) from racist countries of southern Africa is conducted at Rochester University and Lincoln University (a predominantly Negro institution in Pennsylvania).

US policymakers have frequently sought to place emphasis on African education as a substitute for far-reaching indus-

---

[1] Primary education programs alone, however, showed a wide discrepancy of enrollment between different countries, e.g., Niger—3.3%, Mauritania—7%, Guinea—20%, the Ivory Coast—33%, the Congo (Brazzaville)—70%. On the other hand, several countries, in West Africa for example, were considered to have an "oversupply of primary school graduates". *Conference of African States on the Development of Education in Africa*, UN Economic Commission for Africa, Addis Ababa, May 1961, pp. 63-64, 1971.

trial and economic changes. Thus, for example, programs of community development (adult literacy, sanitation, recreation and cooperatives) have been given priority on the assumption that Africa would remain tied to the land, as well as to avoid the "disintegrative" effects of modernization. "Rural uplift," it has been noted, "is in part to prevent social and political upheaval."[1] The promotion of such programs in place of the more dynamic factors has often mirrored the philanthropic, reformist attempt to accommodate to an inert status quo rather than to propel the economy and society forward. Africa, nevertheless, has continued its search for technological advancement in all fields.

With this in view, Washington has been stepping up educational ties with Africa in the fields of science, agriculture and medicine. Thus, a US government team headed by the chairman of the US Atomic Energy Commission[2] visited Morocco, Tunisia, Ethiopia, Kenya, Zaïre and Ghana in mid-1970. Using as a starting point the small nucleus of African administrators, scientists and doctors, most of whom have been educated in the United States or Western Europe, the US task force planned within three years to have the science, agriculture and medical faculty of every African university linked in a "sister college" relationship with its counterpart department or college in an American university. In addition, African institutions were to be placed on the circulation lists of American journals free of charge, a program was to be organized of Americans to spend one year or more in Africa (similar to that pursued by France), the US agricultural extension service was to be adapted to Africa, and exchange and links by US private organizations, universities and individual scientists were to be encouraged.

One of the most ambitious US ideological programs, in the form of education and technical assistance, is the Peace Corps. As grounds for its formation, a biographer records how John F. Kennedy, toward the end of his campaign in Chicago in 1960, remarked that he wanted to demonstrate to the Soviet Union "that a new generation of Americans

---

[1] W. D. Robinson, *Africa Today*, Vol. 14, No. 2, 1967, p. 31.

[2] The choice of the prestigious field of atomic energy is noteworthy. Furthermore, Lovanium, Zaïre, with its 5-kilowatt nuclear research center, has the only nuclear reactor in Africa outside of Egypt and South Africa. See Glenn T. Seaborg, "A Scientific Safari to Africa" in *Science*, August 7, 1970.

has taken over this country...young Americans."[1] Furthermore, that he envied the patriotism of Communist youth of Cuba, where "...each week-end 10,000 teachers go into the countryside to run a campaign against illiteracy".[2]

The Peace Corps, formed on March 1, 1961, with the President's brother-in-law, Sargent Shriver, as its first director, grew from its initial 500 to 5,000 by March 1963 and to 10,000 (roughly its peak) in another year. Although initially Eisenhower had referred to it as a "juvenile experiment" and Nixon as a program for volunteers "who in truth in many instances would be trying to escape the draft", Congressional bi-partisan approval was general and extended from the New Frontier to the ultra-Right, such as Barry Goldwater.

Hardly expected to be of great economic significance, the aim was fundamentally to change the image of profit-seeking US capitalism by trading on the "fund of idealism" of friendly, American volunteer college students. It was too alluring a network, moreover, not to be used by US intelligence agencies for purposes far removed from African development.

Of a total of 10,530 members in June 1966, about one-third (3,421) were in Africa.[3] About four-fifths of these were in teaching and one-tenth in community services, with smaller numbers in agriculture and health services. As might be expected, the preponderant majority were sent to countries where the United States and Britain exerted most influence, and only about one-sixth of the total in French-speaking states. In the latter, Washington in the rivalry for influence with France encountered major political, social and cultural obstacles which eventually forced it, for example, to discontinue English-language instruction in secondary schools.[4]

---

[1] Schlesinger, op. cit., p. 606.

[2] Ibid.

[3] The majority in Nigeria, Ethiopia, Liberia and Tanzania; smaller numbers in Sierra Leone, Malawi, Kenya, Cameroun, Ghana. From *Annual Report of Peace Corps,* June 30, 1966.

[4] In Gabon, the Ivory Coast, Cameroun, Togo, Chad, Niger and Senegal. *Herald Tribune,* February 6, 1967. French language and cultural ties alone extend to dozens of organizations, poets are named ministers, etc. In 1968, there were in the French-speaking states some 28,000 French teachers (almost one-half in secondary or higher education) and 13,000 advisers. *Afrika Heute,* No. 2 and 3, 1969.

The Peace Corps declined in other African countries for various reasons. Relatively minor programs were discontinued: in Guinea in November 1966 in retaliation for Guinean officials being forcibly taken off the Pan-American lines in Accra on their way to an OAU conference in Addis Ababa; Mauritania halted its program in June 1967 on the basis of US involvement in the Israeli-Arab war; Gabon discontinued in December 1967. Major programs were reduced by Tanzania (the first African country to request Corpsmen) from 394 in 1966, to 8 by 1967 and eliminated in 1969, in protest against US foreign policy in Southeast Asia and southern Africa; Ethiopia, which had 280 members in 1962, who made up a third of all secondary school faculties by 1966, had to curtail its program thereafter because of student strikes and demonstrations against US meddling in the country's internal affairs. The total number in Africa had dropped to 2,639 by March 1969, according to Jack H. Vaugn, retiring director of the Peace Corps, because of "the war in Vietnam" and "the program was too large".[1]

Clearly, a number of factors have been involved in the failure of this vast ideological program to achieve more than modest results. The early advantages of playing on American anti-colonial traditions, while the United States was not a major colonial power, soon petered out in the face of US active neocolonial policies. Anti-Communism was losing much of its persuasiveness in view of its baselessness[2] and Socialist policies, which conform to African interests, e.g., assistance without strings. The high US industrial level, the lure of its automobile economy, the enthusiasm of its volunteers in the early stages, continued to be impressive but far from obscured such basic flaws in American capitalist society as US racialism at home[3] and abroad.

---

[1] *Herald Tribune,* July 4, 1969.

[2] "The Russians," according to President Obote of Uganda, "do not have a naval base anywhere in the Indian Ocean nor in the Atlantic Ocean. Nevertheless, Africa is being told that the Russian Navy is threatening certain sea lanes in southern Africa and the most reliable policeman to contain the Russians is the racialist government of South Africa which needs weapons." Speech at Peoples' Conference at Mbale, December 20, 1970. *Morning Star,* December 21, 1970.

[3] At the closing sessions of the UN General Assembly's Fifth (Budgetary) Committee, complaints were made by delegates from Iraq, Trinidad

The Peace Corps itself has brought more advertisement for American capitalism and CIA activity or meddling in internal affairs than technical assistance. As a result of these drawbacks, Nixon's new Peace Corps director, Joseph Blatchford, in mid-1969 was urging emphasis on technically trained personnel rather than liberal-arts graduates, as formerly. It was doubtful, however, whether the new director, who had been the founder of a "privately financed Peace Corps" in Latin America in 1960 ("some of whose funds allegedly came from the CIA")[1] and a proposed larger carrot of technical assistance, could be more successful in obscuring the contradictions between US imperialist and African national interests.

5. MILITARY ASPECTS

US INFLUENCES SOUTH OF SAHARA

Far from being separate from the political, economic, social and ideological spheres, the military is a category which is closely connected, both as cause and effect, and interwoven with the productive forces and social relations.[2] Nevertheless, the army also has enjoyed a relatively independent role in certain societies and periods, and has constituted a decisive force in the state especially in counterposition to the unorganized masses.

In the distorted socio-economic formations in Africa, with corresponding stunted evolution of modern major

---

and Tobago, Libya, the Ivory Coast, Guyana, and Cuba against US racial discrimination as practised either officially in the South, or socially as it is in New York. Press reports, December 1970.

[1] *Newsweek*, June 2, 1969.

[2] Historically, the army has been important for economic development —the framework in which the ancients first developed a complete wage system, as well as an attached guild system of artisans. The special value of metals and use as money appears to have been based on their military significance. Division of labor within one branch was also first carried out in armies. See Letter of Marx to Engels, September 25, 1857. In contemporary North Africa and the Middle East, the army has more than once been a key force in the drive for economic and social modernization and national independence.

classes—bourgeoisie and proletariat, it is not surprising that the less homogeneous intermediate social strata occupy major positions in the new state apparatus, including the army. Their role has varied from country to country, making it difficult to find a common internal denominator. Much, indeed, has depended on the course of postwar developments, particularly in connection with the anti-imperialist struggle, first for national liberation and then, complete independence in all spheres.

The dominant external (with internal ramifications) factor in the military sphere has been imperialism, initially through imposed colonial relations, including either the professional armies created, e.g., by the British in East and West Africa, or through the assimilation of Africans into their own armies as did the French. Comprising a small mercenary force aimed at suppressing internal opposition, the military was a product of the metropole—officered, trained and equipped. Both the ethnic composition and structure were designed to foster internecine strife—in Morocco, mountaineer Berbers were recruited against the Arab urban population; in Nigeria, about four-fifths of the non-commissioned officers were from the northern regions (Mohammedan) as opposed to the southern Christians; in Jordan, Bedouins were recruited rather than Arab villagers. Large local forces were generally not required by the colonial powers, who could and did draw upon their world resources, contingents stationed in neighboring countries or reserves in the metropole itself, e.g., against Egypt, Algeria, the Congo (Leopoldville).

Following independence, the colonial powers did not hurry to evacuate their troops from the young states, but on one or more pretexts military ties were continued through agreements[1] (often made a condition for, and thereby an infringement of, sovereignty), bases, training and equipment (including spare parts). The armies left behind were

---

[1] In 1966, 14 of 36 OAU states still had formal military pacts (unpublished) with imperalist powers—12 with France, Liberia with the United States, Libya with Britain. By 1971, the United States had expanded its bilateral military agreements in Subsahara to include: Zaïre, Ethiopia, Liberia, South Africa, Dahomey, Guinea, Mali and Senegal. (*The Military Balance, 1970-71,* Institute for Strategic Studies, p. 47). In addition, there were military assistance and other agreements kept secret, but some of which have later come to light, e.g., with Morocco.

small and weak by international standards.[1] Of some 400,000 in the armed forces in the mid-sixties, an estimated 250,000 were in North Africa, 34,000 in Ethiopia, and 30,000 at one time in the Congo.[2] Ghana, with a population of 7 million had an army of 15,000, and Nigeria (55 million population)—8,000 (before the war).[3]

A typical army south of Sahara consisted of about 2,000 soldiers, lightly armed, with practically no ability to deploy aircraft or ships.[4] (Moreover, only Egypt—and at the other pole, South Africa—had its own military industry.) In general, the proportion of expatriate or foreign officers was high, weapons and equipment were obsolescent, a mercenary and anti-popular psychology of a privileged, conservative group prevailed, rather than a national and democratic military tradition. As a result, the former metropoles, which had provided the nucleus of the armed forces of the African states (with the exception of Egypt and Algeria), continued to exert a strong reactionary gravitational pull, either through their immediate presence or ties.

Any effort, therefore, to deal with the African military as an internal category, apart from these relationships, cannot be very fruitful. And bourgeois specialists in military affairs, like W. Gutteridge, throw up their hands sometimes with the observation that an attempt to classify coups is "a kind of game".[5] The play of social forces, crises and coups in the new states, which have often appeared as a "struggle over the spoils between competing layers of the Power élite",[6] are no less a product of the colonial heritage

---

[1] Stronger in the North, where, from a purely quantitative viewpoint, 0.44% of the population is in the regular military forces (similar to Latin America); as compared to tropical Africa, where it is 0.07%. Of national budgets, 10-20% is for the military, or 2-5% of the GNP. V. McKay, *African Diplomacy,* Praeger, 1966, p. 70.

[2] D. Austin, *Britain and South Africa*, London, 1966, p. 27.

[3] W. Gutteridge, "Why Does an African Army Take Power" in *Africa Report,* October 1970. Somewhat higher figures are given by J. M. Lee in *African Armies and Civil Order,* Praeger, 1969.

[4] Perhaps a "fleet" of one or two patrol boats. *Afrique Actuelle,* July 1968. In black Africa, for example, only Ethiopia possessed jet aircraft (*Africa Today,* April 1968). Also, see M. J. Bell, *Military Assistance to Independent African States,* Institute for Strategic Studies, London, 1964.

[5] Op. cit.

[6] Ruth First, *The Barrel of a Gun,* London, 1970, p. 429. In this study of Nigeria, Ghana and the Sudan the author sees the major internal adversaries as the "civil service-military" versus the "politicians in business".

and the quite important—often decisive—contemporary world influences.

Like the other categories, it would appear that the military, too, is most understandable in relation to the struggle in the three major areas[1]. North Africa, in which Egypt and Algeria, for example, involved in intense and prolonged struggle against imperialism; tropical Africa, with relatively little prolonged mass militant struggle; and the continent's antipode, southern Africa, with its especially fierce colonial and racist suppression. Of the imperialist influences, however, it is the role of the United States with which we are most concerned.

In the period of collapse of the colonial system and rise of new states, the military activities of the United States, meshing with its political ties and economic interests, had expanded beyond its own forces to global proportions,[2] with policies to encompass the arms of its allies and new military programs. In Africa, the US-fostered NATO envisaged continued spheres of responsibility on the part of the ex-colonial powers in the young states. Although each NATO country has its own imperialist objectives, sometimes at variance with those of Washington (e.g., in North Africa), there are no instances of one colonial country using its troops against those of another. Their individual military agreements covering bases, and the officering, training and arming of local troops, were viewed—at least by Washington—as part of an intercontinental or global network, serving individually or jointly either to prop up regimes protecting imperialist interests or to undermine independent-minded governments. And that is essentially how they have been employed. In 1963, for example, the colonial countries still had 17 air and 9 naval bases in Africa. The greatest

---

[1] A categorization made on the basis of the origin of the army: 1) non-colonial armies of formally sovereign states, e.g., Ethiopia (also Turkey, Iran); 2) former colonial armies inherited by national states, e.g., Egypt, the Congo (K)—(also Iraq, Syria, India, Pakistan); 3) armies formed in the process of struggle, e.g., Algeria (also Burma, Indonesia); and 4) those formed after independence, e.g., most of the African states. Г. И. Мирский, *Армия и политика в Азии и Африке (Army and Politics in Asia and Africa)*, M., 1970, pp. 5-6.

[2] "Without the U.S.," declared President Kennedy, "South Vietnam would collapse overnight. Without the U.S., the SEATO alliance would collapse overnight. Without the U.S., there would be no NATO, and gradually Europe would drift into neutralism and indifference." Speech delivered on November 22, 1963.

number of these was maintained by France, with its biggest tropical African base at Dakar (Senegal), and others at Fort Gurot (Mauritania), Fort Lami (Chad), Abidjan (Ivory Coast) and Diego Juarez (Madagascar). France has special agreements (unpublished) with all of its former colonies, except Guinea and Mali, granting it the right to intervene "to maintain public order". Thus French troops intervened in the Mali Federation in August 1960, were dispatched to the Ghana-Togo border in December 1962, and intervened in the Congo (Brazzaville) in August 1963. On the other hand, the French government did not choose to comply where it did not suit its interests, e.g., ex-President Youlou's request for troops was refused and his Government fell, and a request by Modibo Keita also was turned down. In 1966, the series of military coups in Dahomey, Upper Volta and the Central African Republic were led by French Army-trained officers who a decade earlier had participated in the war in Indo-China.

British imperialism has maintained bases at El Adem and Tobruk in oil-rich Libya (until 1970), Freetown (Sierra Leone), Kano (Nigeria), as well as in East Africa, Rhodesia and South Africa. It had in the sixties about 600 military specialists in Africa, planning "defense" measures, training officers and conducting joint manoeuvres by Commonwealth countries. In 1964, the legacy of a British-trained and -officered army was the cause of army disorders in Kenya and Tanganyika which put the newly sovereign states in the vulnerable position of having to call for the troops of the ex-colonial power to help restore order. Subsequently, to prevent a repetition of the humiliating experience, the OAU at Dar-es-Salam decided to call for the organization without delay of national armies composed solely of Africans.

Belgian imperialism has maintained an important air base near Kamina, as well as military specialists and ties with Katanga, which were important in preventing the young Lumumba government from consolidating Congolese self-rule. Immediately following the achievement of independence by the Congo in July 1960, moreover, all indications point to a mutiny having been provoked by Belgium in its own colonial-trained and -officered Congolese army, in conjunction with a Belgian military-supported, separatist Katanga movement. (For this and subsequent US-Belgian rivalry in the Congo, see later.)

Although the colonial countries' forces in the early postwar years constituted the major element of general imperialist military strength in Africa, Washington also exercised its own direct influence through the presence of US troops, bases and military aid programs. The relationship of the latter with respect to US political-military concepts has undergone some modification especially during the sixties, for example, from early agreements mainly logistically conceived to give access to bases, to a later emphasis on indirect political control plus "a military component capable of swaying the local balance of power".[1]

To exercise such control, the US Defense Department, for example, calculates an optimal military budget—desirable size and composition of military forces—of all non-Socialist states as a basis for the US military assistance program.[2] The official military aid totals to Africa, it may be noted, are small as compared to other regions (about one-fifth of that to Latin America, or one-twentieth of that to the Near East and South Asia, and even less of the Far East). But US official aid figures are notoriously understated.[3]

Within Africa, the actual pattern of US military emphasis is also distorted if one relies on official military aid statistics, which show: over one-half of the total going to Ethiopia, and then much smaller amounts to Libya, the

---

[1] G. Liska, *Imperial America—The International Politics of Primacy,* Baltimore, 1967, p. 98. This was related to political feasibility and resulted, in part, from the unanimously adopted OAU decision in 1963 urging its members not to participate in military pacts nor permit bases on their territories. Although in the three years following the decision, not a single pact nor base had disappeared, an open foreign military presence was becoming more embarrassing and had to be concealed behind secret "defensive" agreements.

[2] Testimony of Townsend Hoopes, former assistant to the Secretary of Defense, at *Hearings of the Joint Economic Committee of Congress,* January 5, 1971. Hoopes declared that former Defense Secretary McNamara "politely ignored" such optimal calculations, but that the Nixon Doctrine of a "lower U.S. profile" argues for "a much larger outlay for military assistance". Press reports, January 6-18, 1971.

[3] Senator Proxmire's *Joint Economic Subcommittee Hearings,* for example, brought out, under questioning, that US military aid figures were at least "eight times the amount officially listed in the President's Budget". Furthermore, the transfers made from the Pentagon's "excess" weapons stockpile are calculated at bargain rates, and, in addition, substantial funds under the aegis of Food for Peace are used to purchase US arms. *Newsweek,* January 18, 1971, and other Press reports.

Congo (K) and Liberia. The large category "Other", for instance, conceals significant programs, e.g., with Morocco. The aid figures themselves, moreover, are dwarfed, in some cases, by other programs, not so listed, but which sometimes have an even greater military and overall impact, e.g., in the Congo (K).

Furthermore, account must be taken of US dealings with regional complexes, through its own, allied, or local reactionary forces. Far from being of a supplementary nature, such arms are frequently of major proportions, e.g., US forces in the Mediterranean in conjunction with Israel's role in North Africa and the Middle East; or South Africa's large military budget and Portugal's arms received via NATO. The "local balance of power" may also be swayed by a gamut of US operations from applied social studies,[1] to intervention, and clandestine activities such as coups d'états. These activities lend themselves even less to statistical measurement, e.g., the cost in dollars with respect to impact or results.

The region which unquestionably has attracted the greatest amount of Washington's political attention, been the greatest source of monopoly profit, and scene of most intense military activities has been North Africa, which must be seen as part of the Middle East complex. For their part, the progressive states, which have struck out most militantly on an independent course, particularly Egypt, have constituted the most formidable obstacle to US imperialism in this area. The resultant conflict is of such overshadowing importance that it will be treated separately in some of its aspects in the following chapter.

In the relatively weak tropical African states, the United States has either supplanted or pushed aside its imperialist

[1] The Defense Department, since the late sixties, finances a program partly classified, through American University's Center for Research in Social Science, to study the political, social, economic and cultural roles of military establishments in the "processes of social change". A Chicago team, for example, headed by Morris Janowitz, is studying military élites in East Africa, Egypt, the Middle East and Colombia. (One study, *Social Structure and Revolution* by Jack Bloom has been published as an army document.) Furthermore, the Air Force Office of Scientific Research maintains its own program of applying research to "understand and support the functions of indigenous military élites". Prominent sociologists employed include Seymour M. Lipset and Amos Perlmutter (*Political Functions of Military Elites: North Africa and the Middle East*). See *Guardian*, March 16, 1968.

rivals in a few selected countries, e.g., Ethiopia and the Congo (K). These countries are not only large and important in themselves, but have an influence and implications for US strategy extending far beyond their own borders.

In Ethiopia, with several centuries of formal sovereignty, Britain, France and particularly Italy have penetrated in modern times. Although the last was routed at Adowa in 1896, it, nevertheless, subsequently continued its presence in Eritrea, and fascist Italy's occupation of Ethiopia in October 1935 was ended only in 1941 by combined British and Ethiopian forces. Attracted by the country's strategic location bordering on the Middle East and Arab world, the United States succeeded in replacing British imperialism by the close of 1952 and entrenching itself in this monarchy.

Symbolic of US postwar predominance is the large number of American teachers in high schools and universities, with English as a medium of instruction even in Eritrean towns like Asmara, where Italian is still an informal *lingua franca.* The United States is the country's most important trading partner, coffee being its main export. American embassy and other civilian agencies in Ethiopia make up the largest official representation in any African country.[1] But, undoubtedly, the major area of US penetration of this country, which retains its feudal system and strong military tradition, is through military agreement, base rights and personnel.

The airbase at Kagnew, outside Asmara, the capital of Eritrea, is not only a communications base which claims the largest high-frequency radio-relay station and listening post in the world, but has numerous air runways, and is reportedly used for the deployment of nuclear weapons in accordance with a 20-year agreement signed in Washington in May 1953 which has recently come to light. The base is maintained by 3,500 persons, of whom 1,800 are Americans (accompanied by 1,400 dependents), who make their presence felt.

Ethiopia has received more than one-half of all US official military aid to Africa—about $135 million (1950 to June 1968—later figures are classified),[2] in addition to eco-

---

[1] *The Economist,* July 20, 1968.

[2] H. H. Hovey, *U.S. Military Assistance,* N.Y., 1965, p. 104, *U.S. News and World Report,* February 23, 1970.

nomic aid. The US Military Assistance Advisory Group of some 110 officers and men is reputedly the biggest in Africa, with high-ranking American officers sitting in Ethiopia's Ministry of Defense some "25 yards away from the Chief of Staff".[1] Although Ethiopia and the United States are not formally allies, US strategists claim that Washington can count on using sea and air bases in that country. Massawa is the headquarters for the Ethiopian navy, which is composed of US-built ships. Ethiopian airlines, equipped with Boeing aircraft, have flights connecting Addis Ababa with Robertsfield (Liberia), as well as a number of African and European cities.

The continental and global implications of the US air and communications systems in Ethiopia—with their ability to maintain contact with a worldwide fleet, to photograph airdromes and installations through spy satellites, and to determine approaches to avoid radar dispositions—came to light after June 6, 1967, when they were reportedly of assistance to Israel. The latter, it may be noted, maintains a major military mission in Ethiopia, helping to operate a counter-insurgency program against the Eritrean Liberation Front in the north, and also quietly assisting rebel tribes in the neighboring Sudan.

Ethiopian officers and combat troops, who have been trained and equipped by the United States with modern artillery and aircraft, have been used both in Korean and in Congo military operations (in both of which, the United States has played a leading role). Training, moreover, is regarded as of especial significance under the US leadership training program for African states, which "goes beyond the military assistance training to other countries".[2] The reasoning is quite pragmatic. Because of its level of organization and discipline in countries which are in an embryonic state of nationhood "there will be many occasions during the next decade when the military will take control of some African governments".[3] This applies, in no lesser degree, to the Congo, where the US penetration took place under the much more complicated conditions of a less stable political and military colonial legacy and vis-à-vis a deeper entrenched rival.

---

[1] Ibid.

[2] H. H. Hovey, op. cit., p. 107.

[3] Ibid., p. 110.

When Belgian colonialism was no longer able through repressive measures or long-overdue reforms to hold down a seething Congolese liberation movement in an awakened continent, colonial rule gave way to independence on June 30, 1960.

Immediately thereafter, Belgian neocolonialist strategy aimed a two-pronged attack. *First,* to cripple the new central government by disrupting the *Force Publique,* withdrawing Belgian administrative and technical personnel, and intervening with its own troops. *Secondly,* to pull mineral-rich Katanga out of the young Republic, and to establish it as a seperate state buttressed with Belgian arms, men and money and strengthened ties with colonial and White minority-ruled Africa. Belgium's man in Elizabethville was Moise Tshombe, educated at an American Methodist school, and the son of one of the Congo's few black "millionaires".

In the United States, Tshombe found support among the ultras, the racists of the South and West: Senators Russell and Thurmond, Herbert Hoover and Charles Dirksen, Barry Goldwater and the John Birch Society. As double agent for financial groups in Belgium and the United States and with connections in the Right Catholic hierarchy, Michael Struelens, a Belgian citizen, conducted an hysterical campaign for the Katanga Lobby.

In opposition to the new Belgian (and, secondarily, British and also French) colonialism in the independent Congo, US neocolonialism, which became linked with a UN action, was portrayed as "anti-colonial." Differences between the two powers in political strategy, however, stemmed not merely from economic rivalry but, even more important, from discrepancies between them in relative strength. US imperialism, with its global power and position in the capitalist world, was playing for much bigger stakes—the whole of the Congo.

Throughout the first and second stage of the Congo operation, US foreign policy was primarily concerned with crushing the Congolese national-liberation movement, and only secondarily with subordinating its imperialist partner/rivals. The application of the US postwar concept of filling a "power vacuum" was rather candidly expressed, shortly after the murder of Lumumba, as follows: "If you throw the Belgians out tomorrow . . . there just really wouldn't be anything underneath . . . the problem is to find a way of substituting U.N. strength for the Belgian strength that has been in there before."[1] The US State Department sought to substitute a pliable central government in Leopoldville, which would at least be tolerated by the Afro-Asian nations, whose attention was concentrated on defeating the ex-colonial power seeking to balkanize the Congo through Katanga secession. US imperialism, in executing its *own* political-military strategy, made use of Belgium's intervention designed to cripple the new Republic's central government and, in Katanga, to repress Jason Sendwe's Balubakat, which comprised 40% of the province's population and 1/3 of its territory where the valuable diamond mines in which US monopolies have considerable interests are located.

In the first stage, from July 1960 to February 1961 (assassination of Lumumba on January 17), the US imperialist counter-offensive sought, within the Congo, to divide and suppress the patriotic movement and to decapitate it of its leaders; and, internationally, to isolate it from its world allies.[2] Behind the scenes, US agencies undoubtedly had a direct hand in getting Mobutu,[3] Adoula and Kasavubu to depose and arrest Lumumba. He was initially denied contact with his own people and the world, and then unconscionably handed over, together with Maurice Mpolo, former Youth Minister, and Joseph Okito, Vice-President of the National Senate, to Tshombe, Munongo, Kibwe, Kimbe and Mutaka—to be murdered by his arch-enemies.

The Katangan secessionist strength rested not so much on its army of 8 to 10 thousand men, writes a former Amer-

---

[1] See Testimony of Assistant Secretary Cleveland in "U.N. Operations in the Congo". *Hearings before Subcommittee on International Organizations and Movements,* House Committee on Foreign Affairs, April 13, 1961.

[2] Washington was mainly responsible: for sabotaging the UN resolution of July 1960, which envisaged the use of UN troops to expel Belgian and mercenary forces, for excluding the Soviet Union and other Socialist countries both in New York headquarters and in the Congo from participating in the UN action, for having the representatives of the Socialist states expelled from Leopoldville, and for abusing UN prerogatives to keep assistance from the central government, e.g., closing Congolese airports to "non-U.N. traffic". (See, for example, *To Katanga and Back* by C.C. O'Brien, and *Congo Disaster* by C. Legum.)

[3] See, for example, *CIA—The Inside Story,* by Andrew Tully, N.Y., 1962, pp. 220-27.

ican intelligence officer and official, as on its "officer cadre—two hundred Belgian soldiers of fortune and three hundred or so mercenaries Tshombe had hired mainly in South Africa, Rhodesia and France".[1] Moreover, "one lone Katangan airplane, a Fouga Magister jet, dominated the skies and made the important difference"[2] in preventing the UN forces from quickly reintegrating Katanga into the Congo. In the middle of the fighting in early September 1961, the British refused refuelling privileges in Uganda for Ethiopian fighters which the United Nations had requested. It was instructive that such minor military forces and modern equipment prevented the United Nations from forcing the withdrawal of the Belgian regulars and the mercenaries, although this failure must be seen in conjunction with a US policy looking to a united Congo through reconciliation with Tshombe and Belgium, rather than an all-out victory over colonialism. In this context, it was understandable why the African and Socialist representatives in the United Nations were accusing Hammarskjöld of holding back after initial UN successes and of colonial appeasement.

In the second stage—from February 1961 to July 1964, US neocolonialism, exercising power through its puppets under a UN screen, sought to beguile Congolese patriots into the Leopoldville government, and when this failed resorted to naked force.[3] In contrast, during this period, the US-backed Adoula government reached an accord with Tshombe "for the peaceful reintegration of Katanga into the Congo".[4]

Thus, it is understandable why at the end of December 1961 Under Secretary of State George C. McGhee was declaring that strong anti-colonial speeches (against Struelens, Tshombe and Union Miniere) by Assistant Secretary Mennen Williams and Deputy Assistant Secretary for Public Affairs Carl T. Rowan had not been

---

[1] *To Move a Nation* by Roger Hilsman, N.Y., 1967, p. 251.

[2] Ibid.

[3] Thus, Kasavubu and Adoula in July-August 1961 lured Gizenga as vice-premier, and Sendwe and Gbenye as ministers, into the government. In September, the Adoula-Gizenga accord was repudiated. Illegally, Gizenga was arrested on January 24, 1962, transferred to a camp of Mobutu paratroopers and held until July 27, 1964.

[4] *The Department of State Bulletin*, January 8, 1962. "The Elements of Our Congo Policy", Under Secretary G. Ball.

"cleared at the highest level of the Department".[1] While verbally associating with anti-colonialism, which "seemed so clearly the tide of history", the US government was more basically concerned with securing harmony with its allies, colonialists and racists, and American investors. When Tshombe, by July 1962, showed no inclination to integrate Katanga—for example, by keeping the big plum, the total revenue from UM, while the central government was collecting almost no taxes except in the province of Leopoldville—the US-supported UN Plan for National Reconciliation attempted mild economic coercion in the form of boycott of UM copper and cobalt ores, seizing Katanga assets abroad, and closing rail lines from Katanga to Rhodesia. But these measures did not avail and Africans rapidly became disillusioned with UN actions.

With Adoula losing influence among the Congolese, the US State Department, fearing radicalization of the Leopoldville government, decided on December 11, 1962 to resort to force to end the secession, and announced on December 20 that the United States was sending a military mission there under Lt. General Louis Truman. Although this move was denounced by Soviet Ambassador Zorin and others as "arbitrary unilateral action", Washington succeeded in exploiting its strong global and UN position to take credit for quelling the Katanga secession by January 16, 1963 (after two and a half years), and thereby also to gain dominance over its rivals in the Congo. But the spirit of Lumumba's struggle against colonialism—both old and new—had not been crushed.

Heavy US military[2] and financial "aid" estimated at about $55 million a year, in addition to Belgian funds and an estimated $500 million expended over 4½ years under the auspices of the United Nations (whose members eventually had become disenchanted with this costly and misdirected operation), proved incapable of holding down the Congolese people. In anticipation of UN withdrawal,

---

[1] *To Move a Nation*, op, cit., p. 257.

[2] "For more than a year now," admitted Assistant Secretary Williams, "the U.S. has been providing military equipment, such as ground and air transportation, to help in the training of the Congolese National Army. Our efforts have been linked with those of Belgium, Israel and Italy, who are performing the actual training of the army." *The Department of State Bulletin*, July 13, 1964.

and in view of the broad gains made under Pierre Mulele in Kwilu province in January 1964 and under Gaston Soumialot in Kivu in April, Washington and Brussels made a deal to have Tshombe appointed by President Kasavubu as Prime Minister in place of Adoula who was unable, despite US aid, to cope with the "economic dissatisfaction and opposition to the central government".[1] This marked the third stage of the US-led counter-offensive in conjunction with Belgium after the withdrawal of the United Nations in June.

With Belgium unwilling to cadre the Congolese army with Belgian officers, and the US preferring to drive from the back seat, "the two Governments agreed in Brussels last month that some sort of mercenary force would have to be organized to supplement the Congolese Army, which has virtually collapsed in the face of the rebel assault".[2] The United States, furthermore, urged on the Tshombe régime to appeal to African governments for troops in the hope that this would "at least provide a diplomatic cover for the mercenary operation".[3]

Despite the use of South African, Rhodesian and Belgian mercenaries as shock-troops, operating under air-cover of US B-26 fighter bombers piloted by émigré Cubans,[4] Tshombe's forces were unable to take a number of key urban centers. And rarely were they able to hold territory through which they had passed. Then, preparatory meetings among Harriman, Spaak and Tshombe took place between August and November preliminary to the infamous US-Belgian paratrooper intervention on November 24.[5]

Based essentially on *military considerations* to enable Major Mike Hoare's White mercenary-led Congolese troops to capture Stanleyville and Paulis, US-Belgian-British intervention constituted such flagrant aggression as to require

[1] Ibid.

[2] *The New York Times,* August 25, 1964.

[3] Ibid.

[4] They were "guided by American 'diplomats' and other officials in apparently civilian positions". *The New York Times,* April 27, 1966. The CIA was sponsor, paymaster and director of this "instant air force". Ibid.

[5] Three months before this, and again on the eve of the operation (November 22), the Soviet Union in an official statement warned that the United States and its NATO partners were preparing to intervene in the Congo to crush the patriotic forces and present the world with a *fait accompli.*

a humanitarian-soaked pretext about rescuing threatened White hostages. This was exposed[1] by African leaders, such as Jomo Kenyatta, who played a direct part in the negotiations in Stanleyville. In the United States, Martin Luther King and five other major Black leaders, voicing broad American sentiment, requested President Johnson to halt US intervention and reverse the anti-African policy being conducted by Washington.

The troops involved in the joint US-Belgian paratroop operation, which was successful in achieving its military objective, soon thereafter were compelled to withdraw in the face of blistering political opposition from the African people and world public opinion. A political solution was needed since an anti-popular régime could not even hold the towns its mercenary forces captured, nor garrison nor supply them, much less hold down the entire Congolese people in the villages and bush.

Immediately after the November 24 aggression by NATO powers, a political crisis was precipitated in the United Nations by US insistence that those countries which did not help finance the Congo operation (amounting to more than the UN regular budget of $100 million annually) lose their vote in the General Assembly. The Soviet Union and a number of other countries had refused to share the costs and thereby, by implication, condone the action directed against Congolese patriots. Washington's intransigence compelled the General Assembly[2] to recess until autumn 1965.

During most of 1965, the United States and its imperialist allies sought to refurbish Tshombe, e.g., by taking him into the OCAM (a move initiated by Houphouet-Boigny, the Ivory Coast) and by announcing, and inviting certain states to observe, the much-heralded Congolese elections. These, it turned out, were either subverted or annulled when held. In the last analysis, attempts to "Africanize" the Leopoldville government without removing the root evil of imperialist influence proved vain.

---

[1] Independent Africa almost unanimously (with the notable exception of the then Nigerian government) condemned this aggression in the United Nations.

[2] The State Department, according to commentators, was not averse to paralyzing the UN General Assembly during this period. Of its 115 members, 35 were from the OAU states and 42 from other developing nations.

Finally, it became apparent that Tshombe, whose government was considered illegal by most of independent Africa, had become too much of a liability and would have to be pulled out of the front-benches. This, in fact, took place on October 18, when he was dismissed by President Kasavubu and replaced for what turned out to be a brief interlude by Evariste Kimba.

US and Belgian rivalry for dominance in the Leopoldville government came to the fore on November 25, when General Mobutu quietly deposed Kasavubu (who had been flirting with OAU anti-Tshombe forces in Accra in October) without bloodshed, made himself President and Colonel Mulamba Prime Minister and instituted military rule for five years.

However attractive it appeared politically to disband the mercenary units as the OAU was advocating, Kinshasa found it militarily inexpedient, for even if small in number they were of critical importance in holding down "rebel activity". However, the latent danger of relying on mercenary forces was again revealed on July 5, 1967, when they revolted against the central government—despite the "international" air-kidnapping of Tshombe a few days earlier on his way back presumably to lead the uprising in the Congo. Although the mercenaries immediately seized Kisangani and Bukavu, the revolt was put down, but not without the help of 3 C-130 transport planes supplied by the United States, as well as several jet fighters—by Ethiopia, and pilots—by Ghana. The military lessons—not least of all, the strength relationship of the African armed with spear, to the better armed and trained mercenary, to the decisive role played there by modern air power—were not lost upon the participants, for they were a variant of what had occurred in 1961 (see earlier). And when opposition political leaders were later eliminated,[1] the possibilities of renewed military confrontation between Katanga-Belgium, on the one hand, and Kinshasa-US on the other, became more remote.

The struggle then passed over largely to the political and

---

[1] Mulele after being promised an amnesty was shot in 1968, Tshombe was not extradited from Algeria and "died of a heart attack" in July 1969, and Justin Bomboko and Victor Nendaka were discharged from ministerial posts after Mobutu had assumed their functions (both were later arrested on October 5, 1971).

economic spheres, where US financial groups such as Bank of America, Rockefeller and Morgan already held strong positions, but hardly enough to pry loose the entrenched Belgian financial-industrial-government interests. Hence, US political and financial influence in Kinshasa naturally focused on control of UM (with headquarters still in Brussels), whose mines were nationalized on January 1, 1967. The government's attorney, Theodore Sorenson, formerly special assistant and speech writer of President Kennedy, is frequently referred to as the legal architect of the Congolese government's settlement with Belgian interests in the UM and of the creation of its successor, Gécomin (La Générale Congolaise des Minerais) in September 1969.

The complicated compromise agreed upon, in essence, gave Kinshasa a controlling position, with 25% of the shares of the new corporation. (Of some 40% of the shares offered to the public, American financial groups could be expected to buy heavily.) The former UM owners were to be compensated with 6% of the value of all copper, cobalt and other minerals produced by Gécomin over a 15-year period (and afterwards, 1% of the value of production as remuneration for technical cooperation). Such payments were guaranteed by entrusting the marketing of Gécomin's output to Société Générale de Minerais (a subsidiary of Société Générale de Belgique—the former Belgian controlling interests).

To gain greater Congolese and OAU support, Kinshasa undertook to satisfy broad anti-colonial sentiment by such actions as paying belated tribute to Lumumba and eliminating the Belgian names of Congolese cities in mid-1966. This Africanization trend was continued with the country's redesignation as the Republic of Zaïre after October 27, 1971 and the adoption of a new flag in November. Furthermore, foreign-sounding names were changed in the course of a broadening campaign in January 1972.

Although the struggle had passed over largely to the political and economic spheres, the Congolese army continued to be the key element of power, with American advisers gradually easing out Belgians, US military missions increasingly in evidence in Kinshasa, and its growing dependence on Washington for training, weapons, equipment, and even pay for troops. The Congolese national army, which numbered 5,000 at independence, rose to 60,000 by the close of 1970, and was expected to be 80,000 by 1973. Air force pilots—some in the United States, others Italian-trained—were learning to fly C-130's. Paratroops trained by Israelis numbering 7,000 in 1970 were to increase to 10,000 by 1973. Airports and control towers were manned by some 300 American technicians. Transportation and communications equipment was furnished by the United States, providing Washington with a vital grip on the country's civil life and national defense.

* * *

The military successes in the Congo of US imperialism, especially its organized joint intervention of November 1964, had significant global implications. A view current in the United Nations shortly thereafter was that the employment of Western arms at Stanleyville had evolved into a State Department thesis holding that the well-timed application of US force could stamp out national-liberation movements—a formula fitting into the strategy of "flexible response" which was applied particularly to Vietnam. This turn of events presaged, according to Presidential advisor W. Rostow, the impending "end of romantic revolution" in the world.

In tropical Africa, Washington's reinforced confidence in the decisive role of the military may not have been unrelated to the succession of coups which took place in the second half of the decade. Moreover, these generally reactionary gains, in turn, undoubtedly entered into Washington's estimate of the colonial and racist régimes' ability to continue to hold back the overdue social changes in southern Africa.

Centuries of repression on the part of small minority ruling classes of overwhelming majorities of African populations, who were unable to redress their grievances neither domestically, nor through political action or economic boycott in the United Nations, had led to armed struggle in the Portuguese colonies in the first half of the decade and in Rhodesia in August 1967. South Africa, apparently under prior military agreements worked out for joint action against the liberation movement, immediately dispatched 500 of its security troops trained in anti-guerrilla tactics, with planes and armored cars, to suppress the Zimbabwe African People's Union (ZAPU) freedom fighters in the Wankie area. The key position and role of general military support assumed by South Africa was later confirmed by Prime Minister John Vorster, when he declared on September 22 that the use of such troops "will continue in any area where South Africa is allowed to fight". By spring, it was conservatively estimated[1] that some 2,700 South African troops, in addition to air and armor, were supplementing the 3,600 Rhodesian regulars. In Mozambique at the same time, two South African battalions[2] had been sent to Tete to operate against Frelimo guerrilla fighters. The latter, despite enormous Portuguese military efforts, already were in control of one-sixth of the country with an administration over one million of the colony's seven million people.

Without South African support, the small racist minority in Rhodesia and obsolescent Portuguese colonialism could hardly have evoked military optimism. The US State Department advisor Vernon McKay, for instance, claimed in the mid-sixties that no African force was a match for white-dominated southern Africa on the basis of military strength: Portugal—60,000[3] troops in Angola and Mozambique; Rhodesia—30,000; and South Africa—120,000 to 250,000.[4] It is instructive that such calculations presumed joint operation or a combination of reactionary forces in which South Africa provided the overwhelming share.

In contrast, to be sure, the African masses in these countries and in the independent states still lack military organization, weapons and training. Of no little relevance in this regard, however, are imperialist efforts to prevent their reinforcement, e.g., the British Labour government's refusal in September 1967 to sell arms to threatened Zambia,[5] Rhodesia's northern neighbor (whose army remained British-officered until January 1971). At the same time, these same policies are justified in the bourgeois press by deprecating black Africa's military strength, with even occasional half-veiled encouragement to South Africa to push

---

[1] *The Economist*, May 10, 1968.

[2] Ibid.

[3] Later figures are larger, e.g., 105,000 is given by *The New York Times* correspondent M. Howe (in *Africa Report*, November 1969).

[4] V. McKay, *African Diplomacy*, Praeger, 1966, pp. 150, 165-70. Calculations provided in A. C. Leiss, *Apartheid and U. N. Collective Measures*, Carnegie, 1965.

[5] Basil Davidson in the *Sun*, November 9, 1967.

its influence northward against Zambia.[1] Condescension toward the Africans extends even to excluding their ability to sustain a large-scale national-liberation war such as in Vietnam, according to *The Economist*, "in view of African military ineptitude, thousands of miles from the nearest Communist source of supply".[2] (The possibility of Western assistance to black Africa is clearly ruled out.)

Voicing analogous US ruling class cynicism, a US organ of big business[3] projects an estimated 50 years of continued white domination. Seeking an historical parallel in ancient times when, it notes, 40,000 Romans held down 1,500,000 Britons for 400 years long before the advent of modern technology, it reveals the traditional minority-oriented class reliance on weaponry and better organized armed forces to repress the masses. Such estimates of military strength, which are influenced by economic and political considerations, in turn, have their effect on the foreign policies of the United States[4] and other imperialist countries, in particular with respect to arms supply to Portugal and South Africa.

Washington's military build-up of Portuguese colonialism, it is of significance, pre-dates its big investments. It is mainly via NATO dating from the early 1950's and, according to the late Dr. Eduardo C. Mondlane, leader of Frente de Libertacas de Mozambique (Frelimo), it amounted to half a billion dollars between 1951 and 1961.[5] American military equipment, reportedly supplied during this period, has included: 50 Republic Thunderjets in 1952, 18 Lockheed Harpoon bombers in 1954, 12 Lockheed Neptune bombers in 1960-61, as well as Skymaster and C-47 Dakota transport planes.[6] Naval vessels, too, have been supplied including

---

[1] "It is untrue," wrote *The Economist*, for example, "that this would necessarily bring other African states' vengeance upon any half-stooge emerging from a half-South African financed coup d'état, because probably nobody would be able to prove the charge; and the OAU has got wearily used to many of its members being the creatures of coup d'état." July 27, 1968.

[2] Ibid.

[3] *The Wall Street Journal*, September 22, 1969.

[4] George Ball, for example, finds the current ostracism of South Africa "unpleasantly self-righteous and futile" in *The Discipline of Power*, New York, 1968.

[5] C. E. Wilson, "Portuguese Africa and the U.S." in *Freedomways*, Vol. 7, No. 3, Summer 1967.

[6] See *Apartheid Axis, the U.S. and South Africa*, W. J. Pomeroy, N.Y., 1971.

8 minesweepers in 1953-55, 4 large minesweepers in 1955, 3 patrol vessels in 1954-55, 5 patrol vessels in 1956-58, 2 frigates loaned and 1 supplied in 1957. Such military and naval equipment, which according to NATO rules is not to be used beyond the borders of its member states, it is frequently asserted, is aimed to protect US global interests, but the enemy against whom it has been employed has turned out to be African liberation movements.

This has been confirmed by them especially since the uprisings in Portuguese Guinea (early 1960's), Angola (March 1961) and Mozambique (September 1964), when such equipment has been turned against partisans on land and sea. During the 1960's moreover, 50 North American Sabre fighters were reportedly supplied, 30 Cessna trainers and several hundred North American Harvard trainers equipped with guns and bomb racks for anti-guerrilla operations. US bilateral military assistance in the past decade has also been extended to the training of some 5,000 Portuguese officers and soldiers in "anti-partisan courses" at Ft. Bragg, North Carolina.

US military assistance, which is of significant proportions both bilaterally[1] and via NATO, nevertheless, does not cover the growing military requirements of an impoverished Portugal, which keeps about 85% of its armed forces fighting and sinking deeper into colonial wars in Africa and devouring a military budget which rose from 27% in 1960 to 45% in 1967. To meet its huge military expenditures, Portugal must increasingly resort to loans with resultant indebtedness and continued commitment to carrying out what cannot be considered a lone haphazard policy. In 1969-71, for example, loans exceeded $300 million, with US agencies and private banks being the largest single source, and the rest provided by Britain, France and West Germany.

A recent revealing instance has been the signing in Brussels on December 10, 1971 of an extension of the original 1944 Azores Agreement for the use of Lajes air and naval bases on Terceira Is. until February 4, 1974, under which the US government would provide $36 million to Portugal and the Export-Import Bank grant loans to the value of

---

[1] US official bilateral military assistance averaged $30-35 million annually in the 60's, but did not include undisclosed sums for "defense support".

$400 million. That the US-Portugal agreement was not a mere payment for bases was pointedly indicated by Portuguese Premier Marcello Caetano in a nationwide broadcast: "We are helping the U.S. to the best of our means and it is right that the U.S. should help us."[1] Five US Senators similarly interpreted it as a broad foreign policy agreement, one which constitutionally required ratification by the US Senate. Furthermore, in an unprecedented action, Representative Charles C. Diggs, Democrat of Michigan, resigned from the official US delegation to the UN General Assembly to protest White House African Policy "to actively assist Portugal in waging wars against black people".[2] He also critically noted that US votes in the United Nations support the South African, Rhodesian and Portuguese position in Africa, which can hardly be regarded as a haphazard policy.

The US official position with respect to the colonial régime in fact has not remained static but has shifted with the tide of national liberation. In the 1950's, the policy of Washington was openly in close alignment with that of Portugal. But with the upswing of the African independence movement in the early sixties, the US delegate in the United Nations was to be found voicing support for self-determination—even if this was considered merely a ceremonial gesture, among others, by leading New Frontiersmen.

Since mid-1964, however, after the defeat of the neighboring Congolese patriots, Washington changed its course in general on the need for concessions to black Africa and specifically embarked on a more "conciliatory" attitude toward colonialism, seeking "to encourage both Portugal and the Africans to come to a workable understanding".[3] By the beginning of the seventies, although the United States in its African Policy Statement still gave verbal support to the right to self-determination of the people of the Portuguese territories this was qualified by the phrase that Washington would "encourage peaceful progress to that goal",[4] and in the same breath that it was endorsing the Declared Policy of Portugal of racial toleration, which "holds genuine hope for

---

[1] *Herald Tribune,* December 18-19, 1971.

[2] Ibid.

[3] Assistant Secretary of State Fredericks.

[4] "U.S. and Africa in the Seventies," *The Department of State Bulletin,* April 20, 1970.

the future".[1] This after the Africans, having been denied freedom for centuries and given up hope of attaining it by peaceful domestic and international political pressure, already had been fighting for it for a number of years—and with tangible successes.

That US policymakers, nevertheless, are banking on the continuation of Portuguese colonialism with Western support has been revealed perhaps by no one more bluntly than former Under Secretary of State Ball: "Given the comparative strength and the effectiveness of the forces available to each side, the Portugal position would seem secure in Angola, although somewhat less so in Mozambique; while week by week the complexion of Portuguese Africa is almost imperceptibly changing as immigrants arrive from the metropole to occupy the lands abandoned by the rebels."[2] For a settlement, he recommends that NATO allies should display sympathetic understanding, since Portugal needs "the precious element of time". Without doubt, this is connected with such plans as the further integration of the Portuguese colonies and Rhodesia through the $375 million Cabora Bassa Dam and hydroelectric power (3.6 million kw.) project, promoted by South Africa since 1966 and the international consortium ZAMCO formed in July 1968, under which a million white settlers are to be brought into an area where a network of mines and factories is planned.

Portuguese colonialism, moreover, has implications for imperialism as well as the independent African states far beyond the southern part of the continent, e.g., military support for breakaway Biafra in an effort to dismember Nigeria and the Portuguese invasion of the progressive Republic of Guinea in November 1970. When the UN Security Council, after an investigating team's report, sharply condemned Portugal for the latter action, the United States abstained and thereby "suffered a serious erosion of credibility with Africa and the Third World".[3] To avoid further diplomatic embarrassment from identification with a NATO ally in the

---

[1] Ibid.

[2] Op. cit. These lands, indeed, have been confiscated or appropriated by the colonial régime.

[3] Editorial in *The New York Times*, December 10, 1970. Ambassador Yost conceded that the United States has no reason to question the fixing of responsibility on Portugal's armed forces, but feared "very far-reaching conclusions". Ibid.

militantly critical United Nations' special committee on colonialism (Committee of 24), the United States and Britain formally withdrew from that body on January 11, 1971. Clearly, they had no intention of dissociating themselves from, much less curtailing political, financial and military support to, their partner under the NATO shield.

The fact that South Africa, as a colonial and imperialist power in its own right, has dominated the interlinked socio-economic system of Southern Africa is not new, nor is its decisive political-military role. However, active Western imperialist, including US military and political support for South Africa is frequently denied or deprecated.

The military build-up of South Africa since Prime Minister McMillan's "winds of change" speech in 1960 has been far from haphazard. The country's military budget has increased six times in as many years.[1] A Permanent Force, or standing army (ground, air and naval forces), which had been relatively low at 7,700 (1961) rose to 17,300 in 1967. In addition, there is a "citizen force" of about 25,000 (1964), plus a Commando (special volunteers) of about 60,000 (1965) giving a total of more than 100,000. The separate police force of 30,000 (1966-67)[2] is mainly for internal use, and a reserve of 20,000[3] constitutes reinforced motorized police patrols and units trained in anti-guerrilla warfare. How is this made possible?

Nothing is more dependent on economic conditions, perhaps, than modern armed forces. As Engels wrote, "Their armaments, composition, organization, tactics and strategy depend above all on the stage reached at the time in production and communications".[4] South Africa's industry and economy in general (as indicated in previous chapters) is closely interconnected with that of the imperialist powers. Moreover, with the funds provided from the country's extremely profitable production, South Africa is enabled to import weapons and technology at the current world level.

---

[1] From 44 million rands (1 rand = $1.40) in 1960-61 to 255 million in 1966-67. See *Military and Police Forces in the Republic of South Africa,* Department of Political and Security Council Affairs, Unit on Apartheid, UN, N.Y., 1967, pp. 1-2.

[2] Ibid., p. 10.

[3] Ibid., p. 11.

[4] F. Engels, "Herr Eugen Dühring's Revolution in Science", (*Anti-Dühring*), M., 1947, p. 249.

Although South Africa claims self-sufficiency in the production of light arms and ammunition, which could hardly suffice to repress overwhelming majorities equally armed, its reliance for superiority is on modern armor, navy and air force, which come entirely—thus far—from the imperialist powers. Britain, as the classical investor, financier and commercial power in South Africa, was also the main military supplier[1] until the UN embargo in 1963, terminating deliveries at the close of 1964. Nevertheless, Britain has significantly continued to maintain the largest foreign military mission in the country.

With world and domestic opinion pressuring Britain to observe the UN arms ban, France managed to slip in relatively quietly, after the conclusion of its colonial war in Algeria, to become the main supplier of South Africa. Beginning in 1963, France sold at least 20 Mirage-111 fighter-bombers, Alouette helicopters, licenses (which are especially useful in imparting know-how) to produce Panar armored cars, 3 Daphne-type submarines, 15 Super Frelon troop-carrier helicopters, and 9 Transall transport aircraft.[2] South Africa purchased Impala jet fighters from Belgium and was to build 400 of the latter, the engine for which was licensed by an Italian company but was originally designed by Rolls Royce, whose engineers are supervising production in South Africa.[3] The United States, whose role is not mainly in the military sphere, nevertheless has sold Lockheed transport aircraft amounting to about $35 million annually,[4] and licenses to produce light planes (which can be used for anti-guerrilla warfare). Aircraft sales of France in the period 1960-68 were estimated at over $300 million,[5] making it the country's third biggest customer after Israel and the United States. The even larger sales of over $400 million are forecast for the 1970-75 period.

In its rivalry with France for the lucrative arms trade

---

[1] Including naval frigates, jet planes, armored cars, aircraft. *The Military Balance*, for corresponding years, The Institute for Strategic Studies, London.

[2] Chester Croker, "Military Aid to Africa South of the Sahara" in *Africa Today*, April-May, 1968.

[3] *The Times*, January 24, 1969.

[4] *Foreign Report*, London, January 15, 1970.

[5] *Le Monde*, 22 juillet, 1970. An estimate of $500 million total arms orders is given in an editorial in *The New York Times*, July 23, 1970.

with South Africa, Britain on more than one occasion has prepared to back-track on the arms embargo imposed by the Labor government on November 17, 1964. A week later, the government announced its decision to sell 16 Buccaneer strike planes, spare parts and radar already contracted for, but would not approve further contracts. However, three years later, the same government was reported split on the issue of renewing sales, with eight planes already on the Hawker Siddeley production lines. The argument that France would sell arms if Britain did not, and that the latter needed the foreign exchange, failed to overcome popular hostility, however, and the decision was shelved. Three years later, a Tory government raised again the issue of supplying arms to the racist régime and after a visit to Washington in mid-December 1970, Prime Minister Heath indicated that President Nixon had expressed "understanding" for the arms move. On the heels of this, on January 5, 1971, moreover, the US government approved the sale to Portugal of two Boeing-707 planes (costing $9.2 million each), useful as troop-ferrying transport—a breach of the 1961 UN embargo and the first such government-to-government deal (rather than via NATO).

The British government's determination to renew arms sales to South Africa almost split[1] the Commonwealth Conference of prime ministers assembled in Singapore in mid-January 1971. However profitable may be such trade for Britain or its Cabinet Ministers, apparently more is involved than either mere invidiousness of France[2] or an "obligation"

---

[1] Prime Minister Heath claimed Australian and New Zealand support. Ghana, Malawi and the Ivory Coast favored a "dialogue" with South Africa. Kenya appeared to vacillate. Canada, with an eye to its trade with black Africa, opposed the British arms move. Nigeria and India hinted at restricting economic relations with Britain. President Nyerere in London in early October said: "... if arms are sold we will have to question our role in the Commonwealth." President Kaunda on January 11 spoke of "far-reaching consequences", and a number of British projects in Zambia had been shelved and in early January fifteen army and two air force officers—all British—were dismissed. President Obote on January 8 made the only categoric statement: "... if Britain sells arms to South Africa, we would most regrettably leave the Commonwealth." Press reports, January 1971.

[2] France can do it, explained *The Economist*, because her links with "15 balkanized African ex-colonies are not the same as British special links with 800 million brown and black people". January 16, 1971.

to fulfill the Simonstown agreement.[1] The chief argument advanced (on the basis of the threadbare Communist bogey) is the need for Britain's helping South Africa to secure the maritime route around the Cape (every route, to be sure, being a "life-line"). Significantly, this implies the necessity for breaking the UN arms ban as a measure preparatory to the political rehabilitation of the racist régime.

But, political rehabilitation of the apartheid régime has an even greater global than continental context and thus the US role is quite significant. The US position has undergone some modifications since the early sixties when the US representative joined in what Washington considered a "ceremonial condemnation of apartheid" (admonitions of South Africa, declarations that its racial policy was "repugnant" etc.) in the United Nations, but also of watering down African proposals for effective action (e.g., describing the situation as "seriously disturbing" instead of "threatening" international peace and security). The latter would have given the Security Council mandatory rather than recommendatory force,[2] with a resultant increase in international pressure which might have achieved its ends.

The question of an arms embargo, for example, has inextricable global political implications which have influenced the US position on this question. Thus, in the Security Council debate on arms sales in 1963, while Britain limited its opposition to sales which "could be used to enforce apartheid", the United States went a step further in announcing no arms sales as of January 1964. This was a political decision[3] on the part of President Kennedy, which registered as a positive —if half-way—measure with Africa, even though the United States itself was only a minor supplier. At that time, however, African and Socialist states were insisting on an arms embargo and boycott of goods, without loopholes, which Washington steadfastly opposed. In March 1969, shortly after President

---

[1] Provided for the handing over of the Simonstown naval base from Britain to South Africa, whose navy was to be expanded by 20 vessels, worth £18 million, to be built in Britain between 1955 and 1963. Once delivered British commitments ended. (Ibid.)

[2] *United Nations Review*, August-September 1963, pp. 20-24.

[3] Secretary of Defense McNamara, according to Schlesinger (loc. cit.), also viewed it as such in contradistinction, curiously, to State Department officials, who feared losing the "advantages of co-operation with South Africa on a wide range of defense matters".

Nixon took office, Washington's relaxation of its arms embargo by permitting engines for French-built and American-motored (General Electric) Falcon Mystere 20 jets to be sold, reflected a different estimate of the world political scene than in the early years of the decade.

It was also reflected subsequently in the attitude and steps taken with respect to the political rehabilitation of the apartheid state ("policy of contacts", a "dialog" etc.) which were especially marked in the wake of the OAU anti-racist Lusaka Manifesto in the spring of 1969 and then found embodiment in a UN document at the XXIV General Assembly session (South Africa and Portugal voting against). The reasoning was that "if South Africa can establish strong commercial links with black Africa", wrote *The Wall Street Journal*, "perhaps the black African nations will be more willing to overlook South Africa's domestic racial policies".[1]

South Africa itself has been making strenuous efforts to expand its ties beginning with Malawi,[2] Madagascar, Mauritius, Swaziland and Lesotho, and extending them to West Africa. Although Ghana,[3] Gabon, and the Ivory Coast[4] have been receptive, most African countries like Nigeria—perhaps not unmindful of the role of the colonialists in the Biafran

[1] *The Wall Street Journal*, September 22, 1969.

[2] A trade agreement was signed with labor provisions coming into force in November 1967 under which some 200,000 Malawians work in South Africa (additional 80,000 work in Rhodesia.) [*Times of Zambia*, September 12, 1968; see also "South Africa woos Malawi" by Ndab'ezitha, in *Mayibuye (Freedom)*, bulletin of the African National Congress.] Dr. Banda has given up the policy of boycott and prefers "to try new methods altogether—the method of cooperation" (Johannesburg *Sunday Times*, January 28, 1968). Thus, Portugal is financing a new railway line to the sea for Malawi, and her new capital at Lilongive is being built by South Africa.

[3] The official position of General Ankrah and his successor, Prime Minister Koti Busia, for example, was to continue the economic boycott, but unofficial trade was going on through the Canary Is. and Britain, and South African specialists working in the Ashanti gold mines had their children in specially established schools for whites (Johannesburg *Star*, June 1, 1967). Furthermore, Accra officially favored a "dialog" until the OAU Summit in 1971, when in mid-conference it expediently changed its position. At the 9th Summit at Rabat in 1972, it was totally against. Malagasy also turned against dialog and cancelled its cooperation agreement with South Africa.

[4] President Houphouet-Boigny described the boycott and arms ban as "foolish" and called for a conference to work out steps for discussions with South Africa. *Pravda*, November 13, 1970.

secession, or the subsequent invasion of nearby Guinea—have rejected such maneuvers as Premier Vorster's offer of a non-aggression pact. If it is a small but vocal number of African leaders who are in the forefront in this political struggle, they are largely in countries where Washington, Paris and London exert appreciable influence.

Washington itself attaches so much importance to this struggle that it has been forced—in the face of overwhelming world opinion, to work out an elaborate rationale of US official policy which cannot be barefaced acceptance of apartheid. As Assistant Secretary of State Newsom declared, we "cannot do so and maintain our bona fides with even the moderate African governments".[1] In rejecting support for the liberation struggle, or moderate social reforms as advocated by the World Council of Churches, African Studies Association and organisations concerned that "the liberation movements will find help only from the Communist countries", Washington "finds it difficult to see this path as being either right or effective".[2] Similarly rejected, as might be expected, is the OAU and Socialist-supported position in the United Nations of "isolation" of South Africa diplomatically or curtailment of military or economic relations as being "questionable, even if workable". Then, by elimination of options, the only course left for US foreign policy, according to Assistant Secretary Newsom, is "communications", i.e., that "each side knows better what the other side is talking about. . .greater hope". Clearly, a policy not of barefaced acceptance, but one of shamefaced but cunning political rehabilitation.

This recent US foreign policy line has been implemented by US officials encouraging Congressional Black Caucus members to urge Afro-Americans, especially sportsmen and artists who are paid highest world fees, to break the boycott and perform in South Africa. Along similar lines, Assistant Secretary Newsom proposed the appointment of a black American ambassador to South Africa after his visit in November 1970. African and the Socialist states, however, at the special session of the Security Council held for the first time in Africa

---

[1] "U.S. Options in Southern Africa", Address by Assistant Secretary of State David D. Newsom delivered at Northwestern University, *Congressional Record*, February 26, 1971, p. E 1169.

[2] Ibid.

in February 1972, were not to be swerved from their determination to eliminate colonialism and racism and called on Portugal to cease her colonial war, condemned South Africa for her policy of apartheid, although they were not able to demand of Britain that she abrogate her deal with Southern Rhodesia at the expense of the 5-million Zimbabwe people.

The implications of US monopoly investments and profits and US foreign policy are clearly of critical importance to Africa. If the former provided the long-term economic basis for US foreign policy, apparently the rulers of America have broader political-military class considerations, which even if generally parallel to, are no simple outgrowth of, the former. They are, perhaps, even more closely correlated with a conscious policy of support for reactionary minority ruling classes both on a continental and global scale.

London and Washington foreign policies, which in fact strengthen internally the racist and colonial régimes vis-à-vis their overwhelming black majorities and regionally threaten neighboring independent states such as Zambia and Tanzania, apparently also envisage Pretoria as a political partner in imperialist global strategy. To "fill the vacuum" resulting from British retrenchment east of Suez, for example, the United States began building in March 1971 a $19-million base and communications center (to fly both US and British flags and manned by personnel of both countries) on Diego Garcia (British island in Chagos Archipelago, in the geographical center of the Indian Ocean), which the Tanzanian Government's newspaper declared[1] threatened the whole future of the surrounding area. The base will provide a global link between the Philippines and Ethiopia in the US communications chain, in addition to the US spy satellites and tracking stations in Kenya and Madagascar, as well as the already established radar and communications system in South Africa capable of monitoring the movements of all ships in the South Atlantic and Indian Oceans.

In line with the concept of South Africa as a sergeant-major for a US-British imperialist strategy, the former reportedly is seeking a pact with NATO or its recognition as a connecting link between NATO and SEATO. President

---

[1] The *Standard,* December 18, 1970. "The possibility that South Africa might be allowed to use the new base is additional cause for alarm." (Ibid.)

Nixon, apparently, "has accepted the strategic case though he is not going to make himself unpopular with anybody by saying whether he thinks selling arms to South Africa is the right way of doing it",[1] It has been suggested more openly by others, including General Hans Kruls, former chairman of the Netherlands Joint Chiefs of Staff and then editor of NATO's publication *NATO's Fifteen Nations,* that South Africa should become an "outside member" of that organization.

The US and British concept of South Africa as a junior partner in Africa and the Indian Ocean areas—like that of Israel in the Afro-Arab world—which is but another variant of an alliance directed against national-liberation, working-class and Socialist (all frequently dubbed "Communist") movements, has implications for states such as Bangla Desh and India, as well. For African and world progressive forces, in general, this clearly would imply a joint and principled struggle against not only the predatory exploitation of foreign monopolies and financiers, but no less against the aggressive political-military foreign policies of imperialism.

US, AFRO-ARAB STATES AND MILITARY CONFLICT

GENERAL

As part of the continent of Africa and closely linked with the Arab Middle East[2] (where the imperialist stakes are very high), North Africa (where the progressive states emerged strongest as an anti-imperialist force) has been during the

---

[1] *The Economist,* January 9, 1971. This conservative organ suggests a "political price" be paid by South Africa: "an easing of the banning system, more money for African welfare, the release of a few prisoners."

[2] The terms Near East and Middle East, the latter apparently invented in 1902 by American naval historian A. T. Mahan (see *Foreign Affairs,* July 1960, pp. 665-75) to designate generally a region extending from North Africa to as far east as India, are variously used and interpreted, either separately or together, and sometimes interchangeably. The terms include Egypt, with territory both in Asia and Africa although the "Maghreb", or Arab West (Libya, Tunisia, Algeria and Morocco) is usually dealt with separately. The US Department of State in its official papers changes time and again its definition of the region, leading even specialists sometimes to conclude "it is an amorphous area which cannot be defined". (*The Big Powers and the Present Crisis in the Middle East,* ed. by S. Merlin, New Jersey, 1968, p. 23.)

postwar period one of the world's acute arenas of conflict. Strategic aims and political alignments closely coupled with economic interests have been the motivating forces of imperialism.

The complex of North African and Middle Eastern oil is particularly important since the late 50's because of Washington's fear of a political chain reaction affecting this enormous source of modern wealth. Here the US investment in the middle of the 60's was conservatively estimated to have grown to $4.5 billion (of which about $1 billion was in Saudi Arabia and $500 million in Libya). This was more than the officially acknowledged investment in all of continental Africa. Moreover, by the extraction of one of the world's highest rates of profit from the oil of this region, American monopolists derived approximately one-third of all of their overseas profits. In addition, the fact that Britain and France have large investments and some 60% of Western Europe's oil—for a petrochemical industry in which US monopolists also have a stake—was coming from the Moslem world served to pyramid further US regional into global policies. The inevitable conflict especially with growing Arab national forces has resulted in either preparations for war or open military clashes in most of the postwar years. The stage, however, had been set earlier.

The rival European colonial powers determined the political-military strength pattern in this region in the century before US imperialism began to play an important independent role. This had both regional and inter-continental aspects. Thus, at the turn of the century, the "Eastern question" posed by the imperialist powers involved the decline of the Ottoman Empire of Turkey (the "sick man of Europe") and its replacement by their own form of domination. "For," as Lenin wrote, "any other basis under capitalism for the division of spheres of influence, of interests, of colonies etc. than a calculation of the *strength* of the participants in the division, their general economic, financial, military strength etc. is *in*conceivable" (original emphasis).[1] The major European colonial powers, England and France, succeeded in achieving a dominant position in this area by

[1] V. I. Lenin, "Imperialism, the Highest Stage of Capitalism", in *Selected Works*, Foreign Languages Publishing House, M., 1952, Vol. I, part 2, p. 558.

first checking their common rivals—tsarist Russia, in the latter half of the 19th century, and thereafter the imperial ambitions of Germany, which also was attracted by the Middle Eastern oil resources and strategic position.

The resultant strength pattern prevailed—albeit somewhat modified by the impact of rising new forces—between the two world wars. Thus, after the defeat of Germany in alliance with Turkey in World War I, Britain and France had to retreat from their secret Sykes-Picot agreement of April 1916 to divide between themselves the Arab portions of the Ottoman Empire. For, with the October Revolution, Soviet Russia had repudiated secret treaties and announced their provisions to an attentive world.[1] Nevertheless, the two powers exercised control especially through mandates and treated the countries largely as client states between the two world wars. Although during World War II, Washington continued to look upon the Middle East as a British sphere—even if weakening—of political and military "responsibility", with the exception of Saudi Arabia and Palestine,[2] the United States made its presence felt by supplementally establishing bases in Libya, Egypt, Saudi Arabia and Iran. The eroded position of metropolitan France was reflected in her dwindling influence in Syria and Lebanon in the latter years of the war, and by 1946 all French troops were withdrawn from these mandates.

The world-wide disintegration of the colonial system after the Second World War gave the North African states a new role in the emergent Afro-Asian world. But while, on the one hand, their independent course was supported by the Soviet Union and other socialist states, on the other hand, the declining imperialist hold of Britain and France was augmented or replaced by a more active US foreign policy of regional penetration and intervention, as well as global encirclement of the USSR.

US direct, large-scale power involvement in this region is a post-World War II phenomenon. The cold war against "Communism", however, could scarcely conceal a major thrust against anti-imperialist Arab nationalism, leaning to various

---

[1] See A. Williams, *Britain and France in the Middle East and North Africa, 1914-67*, London, 1968, Chapter I.

[2] See J. C. Hurewitz, *Middle East Dilemmas, Background of US Policy*, N.Y., 1953.

degrees upon reactionary forces in Turkey, Israel and the conservative Arab states. In March 1947, after Britain could no longer "shoulder responsibilities" in Turkey and in Greece (where patriots held three-fourths of the territory),[1] the Truman Doctrine trumpeted Washington's takeover bid. Turkey, with her strong militaristic tradition, then became the single country in the Middle East (assuming Greece to be essentially a European state) to identify herself intimately with Washington, and with the creation of NATO was invited to join. This active US policy soon extended to Persia and operated in "complete agreement" with Britain, as Ambassador McGhee declared at Istanbul in November 1949, though "with not too close an association" which would "tarnish the American image".[2]

Israel, for its small size, has been perhaps of unique value to US policymakers not least of all because of its peculiar position in the midst of the Arab states. Shortly after the establishment of the State of Israel in 1948, with the Soviet Union casting its vote for independence from British domination, ambitious leaders in the Zionist movement veered the country's foreign policy onto an expansionist chauvinist course linked closely with British and American imperialism. Since then, first London and then Washington have made use of the Western links and orientation of the new state to use it as a battering ram against progressive Arab states, a means of further penetrating Africa and Asia, and in the cold war against the Soviet Union. This, furthermore, has enjoyed the double advantage of appearing to export the many-centuries-old unsolved Jewish problem in the capitalist world—presumably to be solved in the Middle East, and then screening American imperialism behind the propaganda diversion of nationalist struggle between a small harried state and a hostile Arab environment.

The early 1950's is marked by US attempts to harness militant Arab nationalism in an imperialist-controlled regional framework. However, attempts to build a Middle East Defense Command to include the key Arab country, Egypt, together with Britain, France and Turkey, foundered

---

[1] See *The U.S. and the Middle East*, ed. by C.G. Stevens, N. Y., 1964, Chapter 6.

[2] Bernard Lewis, *The Middle East and the West*, N.Y., 1964, pp. 128-29.

because of Egyptian national opposition.[1] Egypt was suspicious of Turkey and also demanding that Britain give up her base on the west bank of the Suez Canal—then reputedly the biggest foreign base in the world—and leave the Canal zone, which she held by force after October 1951.

Up to the mid-1950's, the United States steadily increased its political and economic penetration in this region, particularly in its oil wealth.[2] Since this was not paralleled by hoped-for success in the formation of what it considered a pivotal bloc in its global network—a broad military Baghdad Pact (based largely upon the extension of previous Turkish bilateral pacts), Washington was forced to continue its emphasis on its own and NATO air bases built in Turkey, Saudi Arabia, Libya and Morocco, and its Mediterranean fleet. However, to counter-balance its failure to win the predominant Arab states, US imperialism sought greater influence in Turkey and Iran (both Moslem but not Arab), increased ties with Israel, and, as the former US Ambassador to Egypt (1961-64) writes, with certain Arab "traditional monarchies whose position was based upon a landlord and merchant élite rather than upon the greater assent of the commonality".[3] Thus, he continues, "American action was often interpreted as directed toward the same objectives as those pursued by Britain and France in the past".[4]

Washington, no less than London, focused on Egypt as the major political and military force threatening imperialism in this region. The national upsurge which had led to the overthrow of King Farouk in July 1952 had given Egypt the opportunity to act in its own national interest, compelled the British in 1953 to agree to self-government for the Sudan and in October of the following year to pledge to evacuate Suez by July 1956. Beyond this framework, moreover, an independent Egypt by virtue of its key position could and did assist national-liberation movements and exert important influence on two continents, most particularly in the Arab and Islamic world. This could not but disturb im-

---

[1] B. Rivlin and J. S. Szyliowicz, *The Contemporary Middle East*, N.Y., 1965.

[2] By 1957, the United States had control of about 2/3 of Middle Eastern concessions and 3/5 of its oil deposits and extraction.

[3] John S. Badeau, *The American Approach to the Arab World*, N. Y., 1968, p. 10.

[4] Ibid.

perialism, Zionism, and those large Arab feudal landowners and capitalists cooperating with foreign rule. Opposing interests were translated into policies which inevitably led to conflict in one or more spheres.

US policymakers, as might have been expected, overestimated their ability to stymie Egypt's efforts to consolidate power during the critical years 1952-55. When the young officers headed by Lt.-Colonel Nasser sought arms and financial aid from the United States, military aid was offered to Egypt but at a high political price—on the condition of permitting an examination of its internal military programs and installations, affiliating itself with the Baghdad Pact of December 28, 1954, and not developing ties with the socialist states. President Nasser declined such infringements upon sovereignty, despite the fact that control by imperialism of the arms market had given imperialism an enormous lever. How critical this monopoly could be was felt during the Palestinian war in 1948, for instance, when unfit weapons supplied to the Egyptian army by those close to Farouk turned the attention of the Free Officers to the link between the common enemies of the Arab people abroad and at home. Moreover, when Israeli forces staged a big raid on Gaza in February 1955 soon after Cairo had refused to join the US-sponsored Baghdad Pact, Egyptian leaders became convinced of the close connection between Washington and Tel Aviv. It was the conviction that US policy in principle opposed the new regime in Egypt which led Cairo to turn to the Socialist states. As a logical consequence, in September 1955, Egypt boldly arranged to purchase arms from Czechoslovakia, with no political strings attached.

The quest for support from the socialist community, which constituted a turning point in the country's struggle for independence, is also sometimes recognized in the West—albeit in the distorted terms of big-power politics. In this sense the new role of the Soviet Union is expressed by one well known regional specialist as "not the result of invasion, nor of infiltration by stealth: the Soviet Union became a Middle East power by invitation".[1]

---

[1] Walter Laqueur, *The Struggle for the Middle East: The Soviet Union and the Middle East 1958-68*, London, 1969. A similar evaluation is given in *Soviet-American Rivalry in the Middle East*, ed. by J. C. Hurewitz, Columbia University, N.Y., 1969.

US and British imperialism again were the main antagonists in a parallel and not unrelated sequence of events, when Egypt was seeking funds for economic development in what has become the classical case of financing the Aswan Dam. After December 17, 1955, when the IBRD had promised $200 million conditional on the United States and Britain providing 56 and 14 million dollars respectively for the construction of the Dam, renewed pressure was applied on Egypt through the medium of the proposed loan to change its political course.[1] But acquiescence was not forthcoming as evidenced by such diplomatic moves as Egypt's recognition of the Chinese People's Republic in early 1956. Imperialism was not slow to reply—first in the economic and then in the military sphere.

Taking the initiative on July 19, 1956, Secretary of State Dulles administered a calculated rebuff to Egypt by provocatively retracting the previous US Aswan Dam offer (Britain and the World Bank followed suit).[2] This, however, proved to be a grave miscalculation in view of the availability of alternative Soviet assistance. On July 26, Egypt nationalized the Suez Canal in a move to secure control of its own waterways and thereby also to obtain greater revenue (up to then only 5% of the total)[3] for its internal development. Washington's response, like that of London, was to freeze Egyptian assets and to seek "international control" of the Canal (e.g., the Dulles Plan of August 16; and the "Committee of Five" and Canal Users Association plan in September) in order to wrest it from Egyptian hands. But these proposals were rejected in rapid order.

In the subsequent Anglo-Franco-Israeli Suez aggression led by Israeli forces on October 29, the imperialist role is instructive, particularly in view of comparable events a decade later. Evidence at the time (and since then amply proved) pointed to the collusion of Britain and France in inciting

---

[1] It was demanded that Egypt not implement beyond a modest level the agreement signed to purchase arms from Czechoslovakia, that the World Bank supervise Egyptian finances, and that Cairo curtail international political activity "unfriendly" to the Western powers. M. Kerr, "Coming to Terms with Nasser" in *International Affairs*, London, January 1967, p. 71.

[2] John S. Badeau, "Development and Diplomacy in the Middle East", *Bulletin of the Atomic Scientists*, May 1966.

[3] Thereafter, the nationalized Canal earned about £E 100 million a year until June 5, 1967.

Israel to attack.[1] On the eve, Washington did not protest troop concentrations, in effect conceding the admissibility of the use of force.[2] President Eisenhower in mid-September apparently had hoped that the threat of applying armed force by Britain and France might be sufficient to intimidate Egypt.

The fact that the actual employment of military force was not excluded from White House policy is revealed in the personal account of the period by President Eisenhower. On July 31, 1956, for example, a few days following the nationalization of the Canal, he wrote to Prime Minister Eden: "We recognize the transcendent worth of the Canal to the free world and the possibility that eventually the use of force might become necessary in order to protect international rights."[3] However, he was hopeful that other "pressures on the Egyptian government" would be effective. But, he added, if "the situation can finally be resolved only by drastic means" a broad effort should be made to convince public opinion that action was "undertaken not merely to protect national or individual investors".[4]

By October 30, after the Israeli attack was launched (but before the Anglo-French landing), President Eisenhower, completely oblivious of the plight of the victim of aggression but concerned about the alignment of world forces, requested Eden for "some way of concerting our ideas and plans",[5]

---

[1] Thus, Anthony Nutting, Minister of State for Foreign Affairs, under the then Prime Minister Anthony Eden, has revealed (in his book *No End of a Lesson,* London, 1967) that the cabinet as a whole was privy to an *international conspiracy.* "Even at that time," commented the *New Statesman,* "many (including this journal) believed that Britain and France were in collusion with the Israelis, though none then suspected that we had actually egged them on to invade Egypt. Since 1956 the evidence of collusion has accumulated to the point where it has become irrefutable." Foreign Secretary Selwyn Lloyd apparently had a secret meeting in a villa outside Paris with Moshe Dayan and the French. They agreed that the Israelis would attack Egypt and then Britain and France would intervene, calling for the withdrawal of the combatants to within 10 miles of either side of the Suez Canal. A tripartite treaty, which the Israeli representative insisted upon, was signed at Sèvres on 23-24 October, according to Christian Pineau, then French Foreign Minister, and published in *Suez Ten Years After,* ed. by A. Moncrieff, London, 1967. See also A. Williams, op. cit.

[2] *Pravda,* September 16, 1956. See also *American Expansion in the Arab Countries* (in Russ.), Institute of Asian Peoples, M., 1961, pp. 72-73.

[3] D. W. Eisenhower, *The White House Years, 1956-61,* N.Y., p. 664.

[4] Ibid.

[5] Op. cit., pp. 678-79.

in the hope of avoiding a split between the two powers on the question of tactics.

When this failed and the Anglo-French attack took place, however (although Washington had not been a mere passive observer in the steps leading to the triple aggression) the United States faced with a rising tide of world opinion was compelled by November 1 to dissociate itself from it, although without censuring it, and to support the UN action to halt the invasion.

The reasons for this US decision, which the Arab world, according to the Director of the Middle East Institute of Columbia University, considered a "great exception"[1] in Washington's policy, were complex but several motivating factors stand out. Whereas Britain and France had their eyes glued to concrete interests such as regaining control of the Canal (and, if possible, overthrowing Nasser), US imperialism through its global view saw an "ill-conceived and ineptly mounted Anglo-French military action"[2] which "was possible a week after nationalization" but not "three months later when the affair had become a *cause célèbre*"[3] in which various world forces had time to align. These, moreover, were by no means merely military forces.

Looking to the maintenance and expansion of US imperialist political-economic interests in the entire region, Washington faced the dilemma of either support for an outmoded rival colonialism, with methods of gunboat diplomacy, or adaptation to new forms. It chose the latter. This, incidentally, Britain had done already in South Asia, and after its catastrophe in Suez was to apply broadly to Africa. France still needed the full lesson of Algeria to make the transition.[4] Political factors for Washington in this instance outweighed the seduction of military intervention, which offered some prospect of immediate gain, but considerable long-term losses. From an objective point of view, the force of Arab nationalism lay not merely in Nasser's vigorous leadership (for which there was no ready substitute) but also in the demand

---

[1] John S. Badeau, *The American Approach to the Arab World*, p. 10.

[2] Op. cit., p. 5.

[3] Op. cit., p. 7.

[4] In a letter to Churchill on November 27, 1956. President Eisenhower expressed fear of "resentment that, within the Arab states, would result in a long and dreary guerrilla warfare." Op. cit., p. 680.

for the removal at the very least of visible forms of foreign occupation and rule. There was also the global danger that US backing would present a solid imperialist front, polarizing against it the underdeveloped nations and Socialist states. Apparently, Washington also was not unimpressed by the position taken by the Socialist world.[1]

Washington did not err in its estimate of the high political losses incurred by the participants in direct military attack. By the end of November, the UN was calling for the withdrawal of the aggressor's troops from occupied territory, which Israel did not agree to until March 1, 1957. In the interim, the Soviet Union clearly spelled out in its Note to the Western powers of February 12, 1957, basic principles for a Middle East peace, which included the right to one's natural resources, no military alignments, liquidation of bases and withdrawal of troops, and joint refusal to supply arms. The Western powers, however, showed no interest in, for example, replacing the 1950 Tripartite Declaration by a four-power declaration to include the Soviet Union, which could have provided the basis for a more durable political settlement.

From the imperialist point of view, however, the "vacuum" which had developed as a result of the defeat of Britain and France now would be filled by the United States as "guardian of law and order" in the Middle East.[2] This had as its political expression the Eisenhower Doctrine of January 5, 1957,[3] and the change of guards was announced by the military demonstration of strength of the Sixth Fleet at Beirut in early 1957.

The strong ties of the United States to Israel had its negative aspects in hampering the former's efforts to retain

---

[1] On November 10-11, citizens of the Soviet Union, for example, declared their readiness to volunteer in support of the Egyptian people. "There are those who believe," according to a recent MIT study, "the U.S. might not have pressed its British and French allies so exigently to desist from their attempt to overthrow Nasser in 1956 had it not been for fear that Moscow's threat to intervene might be real." L. P. Bloomfield and A. C. Leiss, *Controlling Small Wars: A Strategy for the 1970s*, N.Y., 1969, p. 397.

[2] *The New York Times*, January 2, 1957.

[3] As approved by joint resolution of Congress on March 9, the President was authorized to use armed forces against, as fictitiously expressed, "any country controlled by international communism". See S. Merlin, op. cit., pp. 158-60.

influence in the Arab world—not least of all in Egypt. Thus, Secretary of State Dulles was compelled by the alignment of world forces and public opinion to support the Arab demands for pressure upon Israel to withdraw troops from occupied territories, which, in turn, permitted the opening of the Suez Canal in March. Its successful functioning under Egyptian control, moreover, despite pressure and blandishments by the United States and Britain, laid to rest the fiction that the Arab could not manage without a foreign overlord—whether old or new. And, although Washington had avoided direct military confrontation with Arab nationalism in Egypt, its newly announced Doctrine could hardly be misinterpreted. The removal of US and British troops from Lebanon and Jordan in 1958, and the withdrawal of Iraq from the Baghdad Pact pointed, as might have been anticipated, to diminishing US influence in the Arab world.

Closely linked with Egypt, Syria was a parallel target of Washington in this region in the mid-1950's, especially after the overthrow of the dictatorship of Shishekli in February 1954. US hostility to the progressive course of the Syrian Government was accentuated by economic geography—the country's role in the Middle East transit trade, and its foreign policy and relations with socialist countries. Following Syria's rejection of US military aid with political conditions, Washington went over to a campaign of slander and provocation at the close of 1954. In March 1955, suspicious provocations took place on the Turkish-Syrian border, followed by US demonstrations of strength in the Eastern Mediterranean, and Israeli troop concentrations on the Syrian border. In the latter half of 1955, provocations on the part of Turkey, Iraq and Israel took place, and in 1956 Syria was accused of disturbing the peace of its neighbors. It was largely the imperialist-inspired threats to the common interests and progressive course of Egypt and Syria which impelled them after the Suez aggression to unite to form the UAR (until 1961).

Since the Suez crisis, the United States, occupying a predominant position in the area similar to that of Britain before the war, has focused its main effort against the progressive states. To avoid becoming involved in direct intervention, the United States has supplied military arms, equipment and training to Turkey, Israel and the Arab monarchies, both directly and indirectly through its military ties. As the major imperialist power behind Israel, the United States nevertheless has also, like Britain, sought to avoid the role of "primary supplier" of arms and thereby the danger of antagonizing and uniting the Arab world against it.[1] Instead, Washington has encouraged the Federal Republic of Germany (FRG) to bear this onus, which brought most of the Arab League countries in conflict with Bonn, for example, in early 1965. However, when Bonn reacted by cancelling its arms shipments to Israel, Ambassador Harriman promised her US arms instead.[2] To soften up the Arab world, however, Washington also promised additional sales or grants of arms to Saudi Arabia, Jordan (which purchased weapons from Britain with US grants since 1957) and possibly Lebanon and Iraq.[3] It was clear against whom such arms were directed. In Yemen, particularly, Saudi Arabian and Royalist forces were engaged in hostilities against Republican and supporting UAR troops.

MILITARY BUILD-UP ON THE EVE OF AGGRESSION

Since the mid-sixties and especially since February 1966, when the Left wing of the Baath party came to power in Syria, both Israel and Washington began concentrating greater efforts on an Israeli military build-up.[4] Not everyone took at face value the State Department's apology for the United States becoming a direct major supplier with its frequent assertion that these arms would "correct the imbalance" in the region and would tend to damp down the arms race. Three months later the Soviet Government warned

---

[1] In the 1950's, for example, when Israel was drawing up its plans for the Sinai attack, a request was made to France for "100 tanks (Super Shermans), 300 half-track vehicles, 50 tank transporters". Most of this was supplied in October 1956. (M. Dayan, *Diary of the Suez Campaign*, New York, 1966, pp. 30, 34). Similarly, after 1956, France was Israel's major supplier of combat aircraft while Britain and the FRG were suppliers of armor and ground equipment. In 1962, the United States became a significant supplier of surface-to-air missiles. See Bloomfield and Leiss, op, cit., pp. 331-40.

[2] In 1965, 200 M-48 Patton tanks were to be supplied by the FRG under US agreement. *Time*, February 25, 1966.

[3] *The New York Times*, April 14 and 29, 1965.

[4] Thus, Washington revealed in February that Israel was receiving some 36 Patton tanks; and in May, that Israel had been promised 30 Skyhawk attack bombers (*The New York Times*, April 3, May 20 and 21). Britain was supplying Centurion tanks and two submarines.

against threats to Syria coming from Israel and reactionary quarters in Jordan and Saudi Arabia with the backing of the United States and Britain.[1]

This warning materialized, moreover, on July 14, when Israel alleging sabotage and road-mining on its territory sent its aircraft on a massive "reprisal raid" into Syria. The latter was clearly not anxious to become involved in military action, and since Syria was the victim of aggression, the Afro-Asian and Socialist states came to her support in the UN Security Council. But, with Israel strong militarily, the United States and Britain sought to buffer her politically. To avoid losing moral position, particularly in the Arab world, they politely expressed their disapproval of the Israeli action but at the same time abstained from voting.[2] This contributed to the Security Council's inability to muster the two-thirds majority required even to pass a resolution critical of Israel, much less to take any action.

By the autumn, however, in an effort to make up for lost moral position and adverse world opinion by seizing the propaganda initiative, Israel brought to the Security Council in October a compilation of 61 "terrorist" incidents on the Syrian frontier since 1965. However, through the fog of charges and counter-charges regarding raids and sabotage, one can discern objectively and clearly economic interests, political aims and nationalist policies. One can understand the fears of the imperialists for their oil and strategic position, the Zionist quest for population and territorial expansion, and the feudal and monarchist anxiety for landholdings and political and social position. All these would be jeopardized if the socio-economic changes in the progressive Arab countries were to be imitated in the rest of the Arab world. And like Egypt's nationalization of the Suez Canal in 1956, the emergence of a progressive Syria and stronger UAR a decade later were danger signals to imperialist planners.

The next Israeli aggressive initiative, however, was a large-scale attack unexpectedly against Jordan on November 13, 1966. This was at variance with Washington's political strategy of maintaining close ties with reactionary circles in both countries. It revealed that Zionist chauvinism had

---

[1] TASS statement, May 27, 1966.

[2] *The Department of State Bulletin*, August 29, 1966, pp. 313-17.

its own expansionist aims at the expense of all of its Arab neighbors, which sometimes ran at cross purposes to the more embracive regional and global strategy of American imperialism.

The typical direction of Zionist adventurism against Jordan, however, was corrected in the next few months by once again concentrating on Syria, which had begun to apply pressure on Western oil monopolies for a greater share of the profits from the oil crossing Syrian territory.

SECOND AGGRESSIVE ISRAELI-ARAB WAR

Israel's blitzkrieg launched against the UAR and Syria, as well as Jordan, on June 5, 1967, was its second major war of aggression in little more than a decade. Let us pass over the political-military moves on the eve of the war such as Israel's attack on Jordan in early 1967, which, it was feared in the West, might lead Jordan to closer ties with the progressive Arab states. This was followed by Washington's strong reaction[1] and the subsequent Israeli shift to the Syrian front which brought Tel Aviv into closer alignment with Washington's strategy. The record shows, that the UAR, preoccupied with its own economic development, was not interested in a conflict with Israel, but was impelled to make a series of moves designed to lessen the latter's threat against Syria. Thus, the UAR request for the removal of the UN Emergency Force on May 18 and the closure of the Gulf of Aqaba on May 24. The Soviet Union, too, even according to writers in the West, was interested in and seeking a peaceful settlement.

The actual launching may very well have not taken place without the imperialist-Zionist line up. And while US imperialist forces did not directly participate, but remained in reserve in the eastern Mediterranean during the six-day war, their presence represented an overshadowing and relevant force.

Although all of the specifics of the US role still have not been revealed—it took 10 years before many details concerning the Suez aggression of 1956 came to light—the most essential facts stand out. The available evidence points to

---

[1] L. Binder, *The Middle East Crisis: Background and Issues*, University of Chicago, June 1967, pp. 23 and following.

the United States having provided the diplomatic and political cover[1] for a surprise attack: urging restraint, proposing a US-UAR exchange of visits of vice-presidents, and cooperation with the UN Secretary General—all of which the UAR actively greeted.

In the military sphere, the United States guaranteed Israel that if the Arabs attacked her territory the US Sixth Fleet would undertake military action for which Prime Minister Eshkol publicly thanked President Johnson. There was no such guarantee of Arab territories, it may be noted. Even if every i is not dotted and t not crossed, such as the exact role of the Sixth Fleet, the military technology supplied Israel on the very eve of the aggression, and the spy ship *Liberty* offshore in supplying information, there appears to be little question about the general outlines. The information supplied for the direction of attack from the West, interference with Arab radar etc., undoubtedly contributed to the surprise achieved by the first Israeli air strike on 25 Arab airdromes, which destroyed or put out of action the latter's aviation and predetermined, to a great extent, the outcome of the war.[2]. Thus, Israel, for a number of reasons which it is not our purpose to examine, could in a blitzkrieg temporarily overcome the long-term Arab advantages of numbers and industrial potential,[3] as well as political-moral position.

Like the Anglo-French-Israeli attack on Egypt in October 1956, this attack too was no mere accidental result of moves and counter-moves on the chessboard between Israel and her Arab neighbors. Such moves may have determined the

---

[1] In a Foreign Policy Association study sympathetic to Washington, the question is raised: "Were American messages to Cairo in the first days of June expressing confidence that Israel was not about to attack a deliberate deception?" The author replies: "Probably not; we do not know." M. H. Kerr, *The Middle East Conflict*, New York, 1968, p. 25. As a matter of fact, the Israeli attack was anticipated by *Al-Ahram* on May 26, op. cit., p. 26. Furthermore, President Nasser reveals, in his July 23 speech, that at the June 2 meeting of the Supreme Command Council, he had indicated he expected an attack in 48-72 hours, i.e., approximately by June 5.

[2] A comparison of Israeli and Arab military strength six months before the war by Hanson W. Baldwin ("Israel against Arabs in a 'Shooting Peace' ") showed the latter with slightly more equipment, but the former having the critical advantages of geography, technical proficiency, élan and unity. *The New York Times*, December 6, 1966.

[3] See Mirsky, op. cit., p. 87.

particular time, place and form of the attack, but the fundamental factors which underlay and led to the war continue to exacerbate present relations and to prevent a resolution of the conflict—Zionist expansionism, oil monopolies and US imperialism, not without the help of small circles of Arab feudalism and reaction.

Presenting it in terms of a nationalist conflict between Israel and the Arab states may describe the actual hostilities but fails to reveal both the critical economic stakes and the imperialist role. Here US monopoly capital's direct interest in this war is comparable with the motives for Anglo-French imperialist instigation in 1956. Similarly, Washington's regional and global aims subsume and dwarf the ambitions and capabilities of the Zionist ultra-nationalists.

Such an assessment of the objective strength relationships involved is important not only for an understanding of why and how the war could take place but also may provide at least the general direction in which a basic solution to a most complex problem is to be sought. For without the patronage and encouragement of imperialism—first largely British and then American—not even a modern technologically equipped state such as Israel, possessing neither the economic base nor the manpower, could maintain a militarized economy[1] or pursue its expansionist foreign policy. And, conversely, to conduct such an adventurous policy, the Zionist leadership has become to all intents and purposes a voluntary-captive of imperialism, seeking to realize its own chauvinist predatory plans under the mantle of parallel, if not always coinciding, imperialist interests.

Why the international oil cartels and Washington have sought as one of their primary aims to topple the Syrian and UAR governments is directly linked to the overall profits from oil and the politics of the progressive Arab states. More immediately, at the close of 1966, Syria's Left government, leading the way to a cut in the profits of the oil cartels, was demanding some $280 million for increased transport payments from the Iraq Petroleum Co. (in which most of the biggest oil monopolies are represented) for the use of a pipeline which carries oil across Syria to the Mediterranean. Increases also were demanded by Lebanon. Then similar

---

[1] With a population of about 2.5 million, the country maintains about 10% of its manpower under arms.

demands were made of American companies transporting oil from Saudi Arabia across Jordan, Lebanon and Syria. Furthermore, Saudi Arabia, Libya and Qatar, together with Iran, were seeking to end the current discount allowance, granted to the big oil corporations for tax purposes, which, it was estimated would have amounted to about $160 million in 1967.[1] By May 1967, it had become clear that American oil monopolies would have to dip substantially into their profits. And the end of this political-economic chain reaction was not in sight.

If Syria, therefore, represented the immediate target of American oil interests, the progressive government of the UAR had long ago been marked as Washington's most powerful political stumbling block in North Africa and the Middle East. Moreover, by 1967 the UAR was making impressive economic progress[2] with the assistance of the Socialist states. This included a prospering Suez Canal, with plans for its widening for the transit of 110,000-ton tankers, the developing and refining of oil, the giant Aswan Dam soon to start paying dividends by increasing total arable land by a third and doubling current electrical capacity, a parallel pipeline from Suez to Port Said to pump oil for giant 300,000-ton tankers to save time by avoiding the longer Cape route.

Several fundamental interacting economic, political and military factors which led to the outbreak of the aggressive war against the UAR and Syria (as well as Jordan) involved: increased US economic penetration, particularly in the oil-producing Arab states, some of the fabulous profits of which were being endangered by anti-imperialist measures; a growing militarization of the economy of Israel, inflated by US donations and investment and leading to that country's greater dependence on and political alignment with Washington; the fostering by the West of nationalism and a military build-up against the progressive Arab states. How did American imperialist and Zionist expansionist political-military strategies converge?

Both Washington and Tel Aviv had similar major targets—

---

[1] *Business Week,* November 12, 1966.

[2] See, for example, a revealing article on the eve of the blitzkrieg from a source not especially friendly to the Nasser government, *Fortune,* May 1967.

the UAR and Syria. Since US policies and actions, both direct and through its blocs such as CENTO, had failed to subordinate the Arab countries and to isolate them from Socialist ties and assistance, Washington had come to place great hopes on Israel to club the resistance of Arab nationalism and anti-imperialism, and to help provide a screen for imperialist aims. From a global point of view, a Middle East crisis also served Washington by distracting attention from its increasingly unpopular aggressive war in Vietnam. For the Zionist leaders, the maintenance of an immigration-inflated permanent armed state[1] on a narrow economic base, leading to domestic economic and financial difficulties,[2] could be justified only by a state of war or near-war. Moreover, the UAR and Syria represented the biggest obstacles to the realization of their territorial expansionist aims: in Syria—the headwaters of the river Jordan; in Jordan—the West Bank (tied with the Palestine refugee problem), part of the city of Jerusalem; in the UAR—the Gaza strip, Aqaba and Suez Canal passage.

In the sphere of ideology, Zionism had long sought to rationalize and resolve these contradictions through an intense nationalism which would set aside centuries of Arab history in favour of claims reaching back to remote biblical times. In the period preceding Israel's "preventive" war, this was generally coupled with an appeal to world opinion to help secure the frontiers of Israel, a small homeland carved out for an age-old persecuted people, which was being threatened by Arab nationalism. Certain irresponsible statements not in accord with the interests or policies of Arab governments assisted the Zionist argument by denying the right of existence of Israel. Such unrealistic statements confused the imperialist nature of the war versus the progressive Arab states by tending to cast the struggle in nationalist terms.

The propaganda line of Washington has meshed with that of Tel Aviv. Having contributed to the regional national antagonisms in the interests of imperalism, US policymakers

---

[1] About 40% of its 1966-67 budget, e.g., was earmarked for arms, and 17% to service loans. See "The Crisis in the Middle East" by Meir Vilner, General Secretary, the Communist Party of Israel, in *People's World*, December 10, 1966.

[2] The country was undergoing an economic depression in 1966—inflation, unemployment, etc.—Ibid.

then have sought to explain the conflict in terms of those resulting animosities and thereby to escape from responsibility. US propaganda at times has tried to give the impression that Middle East oil is not important for the United States since it has other world sources and most of it is exported to other countries, primarily Western Europe. (But the fabulous profitability for the monopolies of this region's oil as compared to others is passed over.) Washington indicates that it must accommodate to "pro-Israel feeling" in the United States, thereby casting the domestic scene, too, in nationalist terms. (But obscured is the fact that the war conformed to US government policy and the dominant economic interests, and was aided by the financial and social links of the American bourgeoisie of Jewish origin, rather than conforming to the interests of the mass of Americans of similar ethnic origin. The latter, who bear the brunt of anti-Semitism, moreover, are encouraged to look for a remedy for racial discrimination in the blind alley of Zionist nationalism, rather than in the only basic solution—working-class internationalism.) In the US-encouraged arms race, every new sale of military equipment is made with the blessings of the State Department and accompanied with the assertion that this will "correct the imbalance" prevailing. (This, in fact, encourages Zionism, which not only converts Israel into an unviable garrison state dependent on American imperialism, but also creates a micro "balance of terror" in the Middle East that can only lead to war, destruction and misery for both Jew and Arab.)

To resolve the Middle East crisis, shortly after the Israeli invasion the binational Communist Party of Israel called for[1] the withdrawal of Israeli forces to the armistice lines, abandonment by Rightist Zionist leaders of their adventurous anti-national policy, disentanglement of Israel from dependence on the imperialist powers, recognition by Israel of the national rights of the Palestinian Arab people, and recognition by the Arab countries of Israel and her national rights, including freedom of sea passage. Such a working-class internationalist approach offered the prospect of a principled long-term solution to a complicated colonial legacy.

---

[1] See Central Committee of the CPI policy statement (based on report of Meir Vilner, Secretary) of June 22, 1967. (*The Worker*, July 16, 1967.)

On a world scale, the basis for a settlement was not provided until almost a half year later after tense debates in the United Nations Security Council which finally achieved unanimity in the Resolution of November 22, 1967.

ARMED HOSTILITIES OR POLITICAL SETTLEMENT

The US role in the aftermath of the six-day war differed from, and was more complex than, either its own or that of the Anglo-French following the Suez aggression. At that particular time, the White House gave the impression, at least, that it had been caught unawares by the triple aggression. The political-military disposition of world forces persuaded Washington to dissociate itself from the attack and press for political settlement. This time, as the implicated imperialist patron of Israel, the United States represented a far stronger power than the Anglo-French combination had been a decade earlier. Moreover, by avoiding direct participation in military operations—perhaps considered the Achilles heel of the London-Paris conspiracy, Washington could maintain an appearance of disinterest—a much more advantageous "low-profile" political posture. To a large extent, the unfolding of the US-Israeli political-military relationship vis-à-vis the Afro-Arab states provides the clue to the dilemma—either a new round of war or political settlement.

The Israeli role, as such, has not been so difficult to decipher, for especially since the end of the six-day war, the words and actions of Right Zionist leaders have confirmed political, territorial and demographic ambitions to which might be ascribed the source of the conflict. Cui bono—who has gained?

Israeli armed forces launched a war—called "preventive" —which put them in possession of the Sinai peninsula up to the Suez Canal, the Gaza strip, the west bank of the Jordan, Jerusalem and the Kuneitra district of Syria. Tel Aviv has since pursued a policy of consolidating and integrating the newly occupied (afterwards called "liberated") territories, e.g., introducing Israeli law, exploiting Egyptian oil on the Sinai peninsula, permitting Israeli settlement, with the ultra-Right urging wider permanent colonization. Zionist and religious leaders of all shades favor annexation, although to various degrees—from Dayan's "all areas" now held, to the Chief Sephardic Rabbi's "religious sanction" with respect to Jerusalem, to the Left-wing Mapam's proposal of retaining only an absolute minimum. On the other hand, the binational Communist Party of Israel, under the leadership of Meir Vilner, is opposed to any annexation of the one-seventh of the Arab lands now occupied.

Right Zionist leaders have continued policies which have led to the flight or expulsion of an additional 700,000 Arabs from their homes since June 1967, bringing the total Palestinian refugees to an estimated 1,400,000.[1] These they are seeking to replace with Jewish immigrants, which can be interpreted only as a policy of permanent expulsion and expansion.

Although Israel and Zionism are not synonymous, the latter is by far the country's predominant political force, driven by an expansionist nationalist philosophy[2] which is self-righteously aggressive. As Prime Minister Golda Meir told Stewart Alsop, "I do not want the Jewish people to be soft, liberal, anti-colonial and anti-militarist because then it will perish."[3] In the flush of a successful blitzkrieg, Zionist officials felt, not surprisingly, that the imposition of a victor's peace would clear up—at least for them—all questions, including those concerning refugees and boundaries. In advocating two-way talks with the Arabs, for example,

---

[1] Differences in estimates diverge as much as 40% from this figure. One estimate is that as a result of the six-day war about 700,000 (comprising about 200,000 from the west bank to the east bank of the Jordan, about 100,000 Syrians from Jawlan District [Golan Heights], and about 350,000 Egyptians from Sinai and the Suez Canal Zone); plus about 500,000 Palestinians who did not flee from Jordan and the Gaza strip and were classified as refugees after the fighting in 1947-48. The number of Arabs under Israeli jurisdiction today has been estimated at over one-third of the population (this includes 300,000 Israeli Arabs). See Don Peretz, "Israel's New Arab Dilemma" in *Middle East Journal*, winter 1968.

[2] In 1946, for example, of a Palestine population of 1,973,000 there were some 608,000 Jews, or nearly one-third, many of whom came as refugees from Nazism rather than as Zionists. The latter, from the very outset, had sought to convert all of Palestine into a Jewish state (first with the support of the Sultan of Turkey and later of British imperialism). Although British rulers used the Arab states in the 1948 war against Israel, the Zionist rulers seized and held more than one-half of the territory allotted by the United Nations to the Arab state in Palestine. Both the 1956 and 1967 Israeli aggressions revealed broad territorial ambitions.

[3] Quoted in *Revue de Défence Nationale,* Paris, October 1969.

Foreign Minister Abba Eban predicted that the Palestine liberation movement "would diminish at the stage of discussions and would disappear at the stage of regulation".[1]

Furthermore, the attitude toward withdrawal from occupied lands, which has been elevated to a question of jeopardizing Israel's security, reveals only minor differences. Eban, for example, has referred to Dayan, who is generally regarded as a hawk, as actually more of a dove, in view of his statement before an American television audience that he "would renounce significant portions of territory for the sake of peace"[2] with Egypt and Jordan. Whether territory would be voluntarily conceded is open to question in view of the nature and aims of Zionism and the influence of its press and propaganda (which has shown, for instance, that some 54% of the Israeli population would not give up an acre of occupied territory).[3] Fundamentally, moreover, the Arabs and world opinion are hardly prepared to concede that Tel Aviv has the moral or legal right to bargain with other people's lands.

In the immediate aftermath of the war, world attention looking to a political regulation of the crisis centered about Israel. But, Tel Aviv refused to adhere to the unanimously adopted Security Council Resolution of November 22, 1967, providing for troop withdrawals, mutual security through recognition of sovereignty and boundaries, settlement of the navigation dispute, and a solution to the refugee problem. Instead, it pursued a strategy of escalated "reprisal" raids so that the Arabs, as Prime Minister Meir has said, "will learn a lesson".

The alliance with imperialism continues to hinge strongly on Tel Aviv's awareness of the former's fear of the potential threat coming from the progressive Arab states to profits from oil and its derivatives, on which their industry is greatly dependent.[4] This may explain another aspect of the Israeli occupation as indicated by Tel Aviv a year after the war.

---

[1] Interview in the *Washington Post*, March 6, 1969.

[2] Ibid.

[3] *Revue de Défence Nationale*, Paris, October 1969.

[4] Thus, petrochemicals link up 40%-45% of French and West German industrial products—from automobiles to detergents and plastics. Furthermore, Western Europe, in which US monopolies have enormous investments, imports about four-fifths of its oil from North Africa and the Middle East.

"Each new day that Israel holds up at Suez, in Jordan and close to Damascus shortens by two days the life of these régimes—at least of Nasser and the Baathists."[1] The same target of Zionist and imperialist policy may also have impelled Prime Minister Meir, in her personal message to President Nixon, to have called upon him "not to repeat the mistake of 1957" in bringing pressure to bear on Israel to evacuate occupied territory.[2] But rather to strengthen the latter's position and wait for international pressure to bring about changes in the governments of certain Arab states.

An important part of such pressure is Tel Aviv's "retaliation" against neighboring Arabs as was graphically illustrated towards the close of 1968. On October 31, for example, Israeli planes raided a bridge and power relay station deep in Upper Egypt (less than 70 miles from the Aswan Dam, or about 500 miles south of Cairo) allegedly because of Egyptian violations of the cease-fire. No US censure followed. Months later it transpired that the Israeli General Staff, according to *The New York Times,* had a list of vital Egyptian targets "to be struck methodically one by one". Israeli hit-and-run raids, according to the *Washington Post*, were "designed to deeply wound Egypt, just as Israel has already wrecked Egypt's two oil refineries, worth $150 million".[3] Two days previous to the raid near Aswan, Israeli planes had struck 37 miles inside Jordan.[4]

International repercussions followed the Israeli attack on the Beirut airdrome on December 28 wrecking 13 commercial liners, which was described by Premier Eshkol as an "act of self-defence". "It is difficult to explain as simple coincidence," wrote *Pravda,* "that the attack on Lebanon was perpetrated the day after completion of negotiations concerning the delivery of 50 American supersonic fighter-bombers to Israel."[5] Moreover, several planes of foreign aviation companies, in particular Pan-American, which were scheduled to come into Beirut at the time of the raid, suspiciously

---

[1] *Davar,* June 7, 1968. Cited in *Daily World,* January 3, 1969.

[2] Agence France Presse and UPI from the UN, May 18, 1969.

[3] Quoted in the *Morning Star*, June 4, 1969.

[4] Tel Aviv policy, according to London observers, felt handicapped by a Jordan government with ties to the West, and preferred to have their conflict firmly polarized in an east-west setting. *The Economist,* December 14, 1968.

[5] *Pravda*, December 31, 1968.

did not arrive. Washington did feel constrained to protest the blatant destruction—but as a threat to world aviation,[1] rather that an act of aggression. It is hard to escape the conclusion that the calculated reign of terror of Tel Aviv was part of a broad strategy designed to impose its victor's terms by keeping the Arabs militarily beaten, psychologically submissive and economically disrupted.

Since the war's end, Washington has been continuing its own "two-tier" policy, whereby Israel, in the first tier, achieved the military victory—although the political objective of bringing down or crippling the progressive Arab governments was not realized—and the United States in the second tier provided the diplomatic screen, economic and military support, and reserve.

By occupying one-seventh of the Arab lands, the aggressors with superior military forces and position apparently felt themselves eminently capable of establishing a new territorial status quo—first de facto and then perhaps de jure. In the UN the ensuing political struggle revolved about censuring aggression, both concretely and in principle (so that it "would not be profitable"), and providing a framework for a settlement and peace. Washington repeatedly opposed any motion to censure or action to secure the withdrawal of Israeli troops. When after months of debate the Security Council Resolution was finally approved, the Johnson Administration did not support efforts to implement it through the Jarring mission, turned its back on de Gaulle's proposal for four-power talks, and rejected the Soviet plan for establishing peace by stages with big-power guarantees.

As regards political-military strategy, Washington, although leaning mainly on Israel in the front-line of military activities, has conducted its own policy toward the Arab countries, for example, by seeking to refine the Israeli policy of escalated "reprisals". As contrasted with Tel Aviv's broad undifferentiated aggression against the Arabs in the occupied and surrounding territories, Washington's more sophisticated strategy is calculated to drive a wedge in the Arab world. The policies of both Washington and Tel Aviv, taken together, have brought no normalization but a state of perpetual hostilities with escalating economic cost.

This has continued to be financed from abroad, mainly

---

[1] *Herald Tribune*, editorial of December 31, 1968.

by the United States, as was the case before the war. Since statistical break-down by source and category of funds is not available, it is difficult to determine exact country and official percentages, but it is apparent that the United States and the Federal Republic of Germany provided the overwhelming share of the foreign inflows estimated variously from $400 million annually to as much as several times that amount before the war.[1] Since the war's end, US, FRG and IBRD funds have increased considerably and are estimated as totalling $1,000 million a year during 1967 through 1975.[2] Indicative of the enormous sums involved was President Nixon's request to Congress in October 1970 for an additional $500 million of aid to Israel.

Such funds alone make possible a military budget to keep the country on a continuous war footing. In fiscal year 1969-70, for example, an inflated budget of $2.3 billion contained a military component officially put at 37%, or 19% of the GNP.[3] Obviously, the economy cannot afford such huge expenditures, which result in deficits,[4] requiring higher taxes, more loans and inflationary deficit spending.[5] Tel Aviv then seeks more funds from Washington "because of heavy investments in armaments",[6] thereby intensifying the voluntary-captive relationship.

---

[1] The lower figure ($7 billion during 1949-66) is given by G. Corm, *Les Finances d'Israel*, published in 1968 by the Beirut Institute for Palestine Studies; *Pravda*, December 12, 1968; and *The Economist*, January 11, 1969. A much higher figure ($500 million a year by the US government alone from 1948 through 1968) is used by the former US chargé d'affaires in Cairo in an article in *The Times*, February 5, 1971.

[2] *Mezhdunarodnaya Zhizn (International Affairs)*, No. 2, 1971. A much higher figure ($800 million in 1970 and $1,500 million in 1971 for the US government alone) is given by former US chargé d'affaires David Nes. Loc. cit.

[3] *Herald Tribune*, January 7, 1969. In the following year, it was estimated at 50% of the budget. *Mezhdunarodnaya Zhizn*, loc. cit. In the US budget, for comparison purposes, an official military component of $81.5 billion is 42% of a $195 billion budget, but only about 9% of the GNP. *Herald Tribune*, January 15, 1969.

[4] $615 million in 1968; $700 million in 1969. *The Economist*, January 11, 1969.

[5] About one-half of the 1969-70 budget, for example, was to be raised from taxes, about 40% from loans (the equivalent of 2½ times the average total receipts during the prewar period), and 10% as deficit spending. *Herald Tribune*, January 7, 1969.

[6] *Herald Tribune*, September 30, 1969. Israel was reportedly seeking additional material to the value of $1 billion in the next five years. *The Economist*, October 4, 1969.

The special political-economic Washington-Tel Aviv relationship which is, by the nature of its aims, also a military one, is tied closely to the purchase of sophisticated weaponry from the big imperialist powers. The role of primary supplier, however, has not necessarily coincided at all times with that of dominant foreign political influence. Thus, although before the June war,[1] the United States had assiduously avoided this direct role (military aid and weapons provided to its NATO partners may be considered as the function of an indirect supplier), it had stepped in when needed to "fill the gap". This occurred on September 26, 1962, when it reversed previous policy by providing short-range Hawk missiles. Furthermore on February 5, 1966, the State Department announced that "over the years" it had sold Patton tanks; and on May 20, 1966—Skyhawks, signifying in effect, that it was going over from so-called defense missiles to jet bombers.

After the war, the United States steadily took over the leading role. Under President Johnson, negotiations begun in the fall of 1967 for the sale of fifty F-4 Phantoms were concluded on October 9, 1968,[2] with delivery promised for 1969-70 (the timetable was later moved up); also provided for was six-month training of pilots in Texas and California. The purchase was to be financed by $200 million in credits, guaranteed by the US government.[3] In 1968, Israel obtained 48 Skyhawk fighter-bombers from the United States, which was "counterbalanced" with the announcement of flight or aircraft-maintenance instruction to be given 250 airmen from Saudi Arabia, Jordan, Lebanon, Libya, Morocco and Tunisia.[4]

The process of the United States replacing France as the No. 1 supplier of aircraft was accelerated with the imposition of a French embargo on arms and spare parts on the day following the bombing of the Beirut airport by Israel. The latter, nevertheless, had accumulated reserves based on record purchases made 1½ years in advance.[5]

---

[1] See *U.S. and West German Aid to Israel,* by Asa'd Abdul Rahman, Research Centre, Palestine Liberation Organization, Beirut, October 1966.

[2] Both Presidential candidates indicated support of the deal during the election campaign.

[3] *The New York Times,* December 28, 1968.

[4] *Herald Tribune,* March 17, 1969.

[5] *Figaro,* February 14, 1969. The delivery of spare parts was thereafter resumed. *The New York Times,* July 4, 1969.

By August, Tel Aviv had turned to Washington on a big-scale, seeking $150 million in jet planes, including about 25 more Phantom F-4's (worth $3-$4 million each) and 80 Skyhawk A-4 fighter-bombers (at about $1 million each).[1] US aircraft were becoming the core of the Israeli airforce as American tanks already were in its armored force.[2] Moreover, planes, missiles and electronic systems were, reportedly, more modern and powerful than those provided by the United States to its NATO and SEATO allies. Americans were also serving in the Israeli armed forces without losing US citizenship,[3] which Cairo cited, in a letter to U Thant, as further proof of Washington support to Israeli aggression and as creating a situation reminiscent of US involvement in Vietnam.

A re-appraisal by Washington of the dynamic forces in this conflict, its options and their consequences, became critical questions with the election of President Nixon in November 1968 and his taking office in January. Previously, President Johnson had decided, on the basis of a National Security Council study, not to force Tel Aviv to disgorge its occupied territories as Eisenhower had done under pressure of world opinion in 1956,[4] but to allow Israel to use its gains to force the Arabs to a settlement.

Three major options were open to the newly elected President, according to a *New York Times* editorial[5]: leave the problem completely in the hands of Gunnar Jarring; without seeking long term regulation, try to defuse periodic explosive outbursts while maintaining a silent recognition of the cease-fire and Israeli occupation; undertake more active diplomatic efforts in search of regulation. With respect to the first option, 16 months of the UN special representative's efforts

---

[1] *Herald Tribune*, August 8, 1969. At the same time, it was provocatively reported that Israel "would like Phantoms equipped with racks suitable to carry atomic weapons", which the United States had previously rejected. Ibid.

[2] Thus, according to one estimate: the United States (300 Patton, 200 Super Sherman), Britain (450 Centurion), France (125 AMX). *The Military Balance 1970,* Institute for Strategic Studies, London, 1971.

[3] *Herald Tribune*, October 18-19, 1969.

[4] Washington at that time, reportedly, had threatened to cut off economic aid and arms supply, as well as to invoke war-time tax regulations on private aid from US organizations. *New Statesman*, April 11, 1969.

[5] *The New York Times*, February 2, 1969.

had brought no visible results. The last course, according to several US officials, did not correspond to Nixon's attitude. If that were the case, Washington could afford to acquiesce to President de Gaulle's initiative for four-power talks as a gesture without serious intent. This, of course, would make itself felt in the talks begun in New York in the early part of the year by the four permanent members of the Security Council, which had the primary responsibility for maintaining world peace and the potential for regulating the crisis. The second option, or some variant of it appeared to be the one Washington finally decided to pursue.

Washington apparently felt it was playing a "winning" game, that is, Tel Aviv could remain master of the situation—within the framework of one-tier (an Israeli-Arab nationalist struggle) and impose a settlement, or keep the Arab states subjected to terror and economic disruption for a long period,[1] or again be victorious in a new edition of the June war.[2] Since this strategy is made possible primarily through US-financed military support (with the political and moral losses to fall mainly on Israel), US leverage could be employed to "resolve" differences with Tel Aviv. Evidence points to the new President, like his predecessor, being "convinced"[3] to pursue this risky course. Thus, in supporting the Phantom deal, he had declared that Israel should be given "a technological military margin to more than offset her hostile neighbors' numerical superiority"[4]. His concept of the "balance of power", apparently, was that it should be kept tipped in Israel's favor against the combined power of all the Arab nations.[5] It has been described even as a policy designed to give Tel Aviv "overwhelming advantage".[6]

---

[1] J. C. Campbell, in his chapter "The Middle East" of a study prepared as a basis for the new Administration's policy, writes of the new territorial status quo that "...it may last another 18 years" (p. 460), and later on—an "indefinite period" (p. 464). See *Agenda for the Nation*, ed. by K. Gordon, The Brookings Institution, Washington, D.C., 1968.

[2] Certainly Israeli leaders have expressed such self-confidence, and as prerequisites that the United States maintain an arms "equilibrium" and bar global rivalry (read: Socialist support to the Arabs) in the region. See, for example, Foreign Minister Eban's speech before the National Press Club, *The New York Times*, March 15, 1969.

[3] Ibid.

[4] *Herald Tribune*, December 1, 1968.

[5] See, for instance, Harry Hopkins, *Egypt, the Crucible*, London, 1969, pp. xxi ff.

[6] Bloomfield and Leiss, op cit., p. 346.

In the second tier, the United States has expanded its own and NATO forces in the Mediterranean. It has built the Sixth Fleet to its second largest (after the Seventh in Vietnamese waters), maintains bases in Spain, Italy, Crete, Greece (ports and airdromes are made available), Malta to North Africa (see later). NATO, which has in southern Europe about 1 million men in its armed forces, more than one thousand planes, and hundreds of ships, reflects Washington's decided preference for joint imperialist action. Since October 1968, NATO has established "Marairmed", with naval aviation headquarters in Naples, for air reconnaissance—a decision apparently taken around 1967[1] (rather than connected, as it is claimed, with "events in Czechoslovakia"). If its major purpose is to act as a military and intelligence support for Tel Aviv, Washington simultaneously is provocatively brandishing a big stick not unaware, as US spokesmen such as Ambassador Yost have pointed out, of the danger that the "two superpowers may be sucked in"[2]—which is, indeed, a real danger. This, perhaps, in the hopes of diminishing Soviet-Arab cooperation—which has not proved realistic.

Washington had good reason to be concerned that the world would see through the game it was playing and that the Arab peoples, in particular, would consolidate against it. For that, in the long-run, could far outweigh the initial six-day military victories against them. In this connection, President-elect Nixon dispatched former Governor Scranton on a mission offering prospects of a new "more balanced" US policy with respect to the Arabs. After inauguration, however, President Nixon spoke more of the necessity for the conflict being regulated by the Arabs and Israelis themselves—thereby also seeking to convey an image of the United States as a power standing apart from the crisis. He also advanced five avenues of US discussion, presumably as an earnest of Washington's peace-seeking intentions. With time, it became more understandable as a dilatory and diversionary tactic. Thus, by March, Nixon was replying defensively to journalists who thought "this Administration was dragging its feet in going into four-power talks".[3] US officials were

---

[1] *Mezhdunarodnaya Zhizn*, February 1969.

[2] Charles W. Yost, "World Order and American Responsibility" in *Foreign Affairs*, October 1968.

[3] *Herald Tribune*, March 6, 1969.

quick to lay the blame for this on Israel's tough position, which Washington allegedly was seeking to soften.

For US imperialism, the critical question of political and moral losses bound up with its role in the Israeli-Arab conflict goes back to the immediate aftermath of the war. The first bitter fruits came with at least part of the Arab world's early recognition that the conflict was not simply Israeli versus Arab, but imperialist against anti-imperialist, and that the mainspring of hostility was US and secondarily British imperialism. Within a few days, six states broke relations with the United States[1] and where possible Britain. From Aden to Tunisia, it was reported,[2] demonstrations took place, ports were closed, oil transport halted, boycotts initiated, and in solidarity Tunisia and the UAR reestablished diplomatic relations. Algeria took over five Esso and Mobil Oil marketing firms.[3] Arab cooperation began to evolve on a broad, hitherto unknown scale, first with the oil embargo for two months and then the September Khartoum agreement (no recognition of Israel, consolidation of Arab strength, oil supply to the West resumed paralleled by a grant of £135 million by Kuwait, Libya, and Saudi Arabia to the UAR and Jordan for losses incurred).

Anti-imperialist interests, however, and political positions taken did not always correspond. Thus, in the UN General Assembly, only one-half of the 38 African states supported the Soviet resolution demanding the immediate withdrawal of Israeli forces from occupied territories, and the OAU would not censure the aggression, which Somalia, Guinea and the Arab states called for.[4] At this stage, not many saw the alignment of forces, as did the Baathist leader Malek el-Amin,[5] as objectively comprising imperialist, Zionist and reactionary Arab forces against not only the progressive Arab

---

[1] The UAR, Syria, Algeria, Iraq, Yemen and the Sudan. *The New York Times,* June 15, 1967. The Lebanon took half-way measures, and no action was taken by Jordan, Kuwait, Saudi Arabia, Tunisia, Libya and Morocco. Ibid.

[2] *Pravda,* June 14, 1967.

[3] *Herald Tribune,* August 31, 1967.

[4] *Izvestia,* August 1, 1967. This mirrored both imperialist influence and Israeli penetration in Africa, e.g., with interests in some 40 mixed companies estimated at $200 million, trade, technical assistance, and military training.

[5] Interview in *Unita* quoted in *Za Rubezhom,* August 4, 1967.

states but also the broad masses of the Arab countries supported by Socialist and other world forces.

Since there appeared no realistic possibility of achieving a settlement in the first tier because of Tel Aviv's uncompromising posture, President Nasser made distinct overtures to persuade Washington to influence Tel Aviv to withdraw from the Suez and to arrange acceptable guarantees with respect to borders and waterways. In fact, the superior political and moral force of the Arab position did help to secure adoption of the Security Council Resolution of November 22, which tended to isolate Tel Aviv because of its intransigent opposition. Nasser, in turn, was able to consolidate his position at home (becoming President, Prime Minister and Secretary-General) and to weather the Abdul-Hakim Amir plot.

In the first year after the war, US prestige suffered a sharp decline, with its strived-for anti-colonial image having disappeared and the United States and Israel more often linked as the new imperialism in this area. However, Washington did make some incursions in the Arab world, although it took until March 1968, for example, before US arms again were being shipped to Jordan. And the resumption of US arms deliveries to Tunisia, Libya and Morocco, according to Drew Middleton, "prompted expressions of friendship that would have been unwise in the weeks immediately after the war".[1] Moreover, with US-influenced states, such as King Feisal's Saudi Arabia, impeding Arab unity and action (e.g., at the Rabat conference) and such diversionary moves as the reactivation of royalist attacks in the Yemen Republic, Washington could still entertain hopes of blunting the edge of Arab antagonism. The latter had resulted, for example, in the expropriation of some of the best US-British oil-rich reserves in Iraq,[2] outright boycotts of US firms (Ford, RCA, Coca Cola), and the replacement of US and British trade by French, Japanese and the Socialist states.

The second year of what has been termed the "shooting" peace was marked by a deepening Arab evaluation of and stronger reaction to the US imperialist role.

---

[1] *The New York Times*, July 17, 1968.

[2] The French were given rights of exploitation in much of the former Iraq Petroleum Co. concessions, and, in December 1967, the Soviet Union agreed to assist in the working of the rich N. Rumeila field. *Newsweek*, July 8, 1968.

Time and events radically changed the assessment. "We have to know our friends from our foes," Nasser told the National Assembly.[1] "Our friends are those who back us and give us weapons. Our foes are those who back our enemy and give it arms. To be more precise, the Soviet Union is our friend and the United States is our enemy."[2] Similarly, Anwar el-Sadat indicated that the United States is "our number one and basic enemy",[3] and that 8,000 military experts from the United States, Britain and the Federal Republic of Germany had served with Israeli forces during the June war. In a wider circle, the Sudanese Minister of State, presiding over a meeting of the Joint Defence Council of 14 Arab League members (except Tunisia), also pinpointed the United States as "enemy Number 1 of the Arabs",[4] and further advised that it was imperative for all Arab states to determine their attitude toward the United States.

It was not unlikely that the dashing of Arab hopes for a changed policy by the Nixon Administration during 1969 had not only lowered American prestige and influence among the Arabs to its lowest point, but also had helped to propel them further to the left. Washington's support, moreover, was for a Tel Aviv government in which the "hawks" had grown even stronger since March 1969, when Mrs. Golda Meir succeeded Mr. Eshkol as Prime Minister, and their continued unwillingness to accept the Security Council Resolution and to negotiate except on victor's terms was changing world opinion against them. In the course of the year, three African revolutions took place—in the Sudan in May, Libya in September and Somalia in October. The Sudan, from the outset, developed close relations with the Arab and Socialist states, and instituted a nationalization program. Libya's anti-imperialist régime immediately requested London and Washington to evacuate their military bases (completed by March and June 1970), required all companies (except oil) to have at least 51% Libyan ownership, displaced European and American managers, teachers and techni-

---

[1] Speech on November 6, 1969.

[2] Ibid. Secretary of State Rogers reacted to the speech as "a setback to peace efforts". *Al Ahram*, straddling, declared that Nasser had not called the United States an enemy of the Arabs, but it had adopted "the position" of an enemy. *Herald Tribune*, November 10, 1969.

[3] *Herald Tribune*, November 11, 1969.

[4] *Herald Tribune*, November 10, 1969.

cians by Arabs, and introduced Arabic as a national language. Colonel Gaddafi's new government cancelled contracts for British arms—originally intended against the UAR, and instead placed orders for French Mirages.

Recognition of the role of US imperialism in this area led to a deeper appreciation of the place of the conflict in the world.[1] The struggle was linked to colonial policy in southern Africa and to the imperialist war in Vietnam. African trade unions called for a unified trade union organization in Africa, demanded that military material not be loaded or delivered to South Africa, Portugal and Israel, and that US troops be withdrawn from Vietnam[2].

The converse and implications of this trend for Washington, which underlie the crudely formulated and superficial "domino theory", had been reflected, for example, in a Western survey on attitudes to US withdrawal of troops from Vietnam: "In Morocco, Tunisia and the Republic of South Africa, a retreat from Vietnam would seriously affect confidence in the U.S. ... In the Middle East, the consequences of a U.S. reverse in Vietnam would be felt hard—and fast. Israel and such pro-Western states as Jordan and Saudi Arabia would be weakened, while Iran and Turkey would be driven to reconsider their dependence on the U.S."[3] Such reactionary fears, which boil down essentially to "defeat breeding defeat", to be sure, are more understandable as those of a section of the ruling classes and their influence on the information media rather than of the general public itself.

Steps were undertaken by Washington and the oil monopolies to halt this dangerous trend in the Arab world. The pro-US régime of Saudi Arabia, in particular, has proved useful in preventing a coordinated front from developing against the US-Israel alignment. In 1969, the spectre of achieving greater Arab unity at a summit conference planned to be held in Rabat was frustrated for months by King Feisal, whose close identification with Washington as Israel's

---

[1] At the Cairo Conference of African Trade Unions, President Nasser pointed to the presence of world trade union representatives as attesting to the fact that the "struggle of the Arab people is an indivisible part of the general struggle". TASS, January 30, 1969.

[2] Conference of Ministers of Labor, the Organization of African Unity, March 10, 1969.

[3] *Newsweek*, November 27, 1967.

main supporter may not have been unrelated to an attempted coup against him in August, which was followed by arrests of hundreds of people, including army and air officers later that month. In September, US oil executives concerned about the Libyan revolution met in Beirut, the Lebanon, and in December with President Nixon (present were David Rockefeller, John J. McCloy and former Secretary of Treasury Robert B. Anderson, with investments in Kuwait and Libya). This was followed by a more "evenhanded" approach in US policy, with Secretary of State Rogers announcing two separate plans for Mid-East settlement—for the UAR and Jordan, on December 8 and 18 respectively. Furthermore, a week later, the long-delayed Rabat conference broke up over the refusal of Saudi Arabia and Kuwait to provide more aid to other Arab states, with the complaint that Libya was not paying enough. King Feisal, apparently, also was not satisfied with Nasser, whose dual attack on Washington-Tel Aviv was complicating the former's position vis-à-vis the Arab people.[1]

Jordan was becoming, perhaps, even more vulnerable domestically as a result of the paradox of its close links with Washington and continued hostilities with Israel. An Arab monarchy, with an oversized military Legion[2] trained and commanded by the British since 1921 and no commensurate national industrial base or economy, Jordan had been meeting this discrepancy up to the June war by foreign technical assistance and military aid from the United States and Britain (and development loans from the West and Kuwait). The King's Western dependence and orientation apparently had been well appreciated when, on June 5, 1967, Prime Minister Eshkol through the UN representative in Jerusalem, Odd Bull, informed Hussein of the Israel attack on the UAR and that if Jordan would stay out of hostilities there would be no repercussions (a strong indication that the progressive UAR and Syria were the targets rather than Jordan). But the UAR had by then an agreement with Hussein and Jor-

---

[1] *The New York Times*, January 4, 1970.

[2] From 1948 to 1956, the numbers rose from 8,000 to 25,000. In 1966, there were about 60,000, mostly of rural origin but about 30% to 40% of Bedouin or tribal, out of a 500,000 economically active (and 2 million total) population. P. J. Vatikiotis, *Politics and the Military in Jordan*, London, 1966, pp. 8-11.

dan was already fighting in Jerusalem and began bombing the Natania base.[1]

If the prospect of an Arab victory had been alluring and led to a new alliance, the monarchy in the aftermath of defeat resumed some of its older ties to the West. Thus, in 1968, as soon as it became feasible, Jordan turned again to the United States and more particularly Britain as its primary arms supplier.[2] Nevertheless, neither imperialist pressures as weapons suppliers nor the blandishments of peace plans (such as the Allon Plan of December 1968 or the Rogers proposal a year later), were able to split Jordan away into signing a separate peace.

This has been largely a consequence of the new resistance forces—political and military—which emerged from the refugees, who increased from about one-third of the pre-1967 Jordanian population to perhaps two-thirds (including many who once again were driven out or fled—this time from West Bank Jordan occupied since the June war). The partisans or fedayeen ("those who sacrifice themselves"—in Arabic) gained influence—establishing their own hospitals, social welfare and tax collection. In occupied West Bank Jordan, too, although organized Palestine resistance existed before the war, its growing overall strength, despite fragmentation into rival organizations, became a powerful force directed against both the Israelis and Hussein because of ties to the United States and Britain. The principal fighting organization, Al Fatah, although founded shortly before 1956, is mainly a post-1967 phenomenon—probably as strong as all the others taken together, with an estimated 5,000 to 15,000 troops.

After the June war and especially the Lebanese events at the close of 1968, the impetus increased for coordinating the various resistance organizations of Palestine refugees scattered in several countries (Jordan, the Lebanon, Syria and the Gaza strip). The largest of these, the Palestine Liberation Organization (PLO), came under the leadership of Yasser Arafat in February 1969. Furthermore, a broad Palestine Resistance Movement was formed to embrace the Palestine

---

[1] For its 22 planes, it may be observed, Jordan had only 16 pilots, the others being in the United States. See King Hussein, *Our War with Israel,* Dar-al-Nachar, Beirut, 1968.

[2] Interview of King Hussein with *Al Ahram* of March 19.

Liberation Organization (its military arm—Al Fatah), the Vanguards of the Popular Liberation War (its armed detachments called As-Saika), the Popular Front for the Liberation of Palestine (PFLP) and others. In the struggle against foreign occupation, a number of partisan organizations have, unfortunately, resorted to adventurist terrorist actions. Progressives, on the other hand, have opposed terror, considering it an instrument of reaction, provocation, or an emotional outburst of oppressed peoples resulting from frustration rather than a weapon of organised mass struggle and revolution.

Terrorist tactics, such as hijacking foreign planes which Al Fatah has censured, not only lowered the prestige of the perpetrators but were used as a pretext by reaction to deliver blows against the partisan movement as a whole. They became the basis, for example, for King Hussein's declaration of martial law in September 1970.

The monarch's subsequent launching of an armed attack against Palestine liberation fighters, however, reunited the resistance movement in the face of a common enemy and threatened to become a civil war. President Nasser criticized the Jordan Army and took steps to mediate, which eventually were successful in getting Yasser Arafat and King Hussein to agree to a cease-fire, together with the resignation of the Jordanian Prime Minister. Washington demonstrated its backing of Hussein during his visit two months later by letting it be known that the United States, in its show of strength in the Mediterranean, had been prepared to intervene if the King tottered in September and by "giving him $30 million and possible further aid".[1]

Although the neighboring Lebanon has not been immediately involved in hostilities, the growing strength of the partisans in the southern part of the country also had aroused anxiety in Washington. The Lebanese Army attacks on refugee camps on October 20, 1969, were related generally to Assistant Secretary of State Joseph Sisco's expression of concern for any infringement of the country's sovereignty "from whatever source" on October 10 and the note of the US Embassy in Beirut with "a broad hint on the presence of anti-Israel commandos in the area."[2] The resignation of

---

[1] *Washington Post*, December 12, 1970 (editorial).
[2] *Herald Tribune*, October 23, 1969.

Premier Rashid Karami, in protest against the Lebanese Army attacks on the refugee camps without his knowledge, followed by mass demonstrations of support for the guerrillas, created a crisis in which both Tel Aviv and Washington appeared tempted to fish in troubled waters.

Under these conditions, the personal efforts of President Nasser, coordinating with Syria and Jordan, brought about a meeting in Cairo of representatives of the Lebanese Army command and the Palestine Liberation Organization which negotiated a settlement. The timely TASS statement,[1] pointing out that "no foreign intervention of a big power could be justified" and that none better than the Arab states themselves "know their own interests and aims", undoubtedly played a part in cooling passions. Furthermore, the Soviet statement indicated that the United States, if really interested in Arab independence and sovereignty as it claimed, could be exerting greater efforts to secure the "withdrawal of Israeli forces from occupied Arab lands and a just solution of the Palestine refugee problem".

President Nixon's Middle East policy revealed itself in the course of the four-power talks, which Washington had reluctantly agreed to and Tel Aviv opposed,[2] the first meeting of which took place in the French UN mission in New York on April 3, 1969. In the State Department's "working document" submitted in the very beginning as the basis for discussion, Washington showed its general support for Israeli territorial ambitions: entire Jerusalem to be under the control of Israel with Jordan given some voice in the life of the city, its emphasis on Israel's determination to hold strategic areas, and its proposal that UN troops be stationed on the Egyptian side of the Suez Canal and the Sinai peninsula. Although few were optimistic about the talks, it was rather widely viewed as a forum to prevent a big-power confrontation. Moreover, to recoup some of the United States political

---

[1] *Pravda,* October 25, 1969.

[2] The Soviet Union, which even in May 1967 had considered the idea "with judicious favor", proposed a concert of the big powers in late autumn 1968 which came to a head in January 1969. France had tirelessly proposed four-power collaboration since the war, the British attitude was ambiguous, and the United States, supporting Tel Aviv's demand for direct Israeli-Arab negotiations, saw "little value in it". P. Windsor, "Super-power Diplomacy: The Price America Will Have to Pay" in *The New Middle East,* London, February 1969.

losses which Scranton had reported in the Arab world,[1] President Nixon was under pressure to show prospects for a new US policy.

In the succeeding months, it became clear that the negotiations in which US backing for aggressors was counterposed to Soviet support for victims of aggression had reached "a blind alley",[2] and no progress for a peace settlement was being made: In substance, the Soviet thrust was to make Israel give up the occupied territory and the resettlement of the refugees; while the US thrust was for Israeli-Arab negotiations in order to impose victor's terms with respect to territory and refugees (Israel and the UAR, for example, were first to agree on "secure and recognized boundaries"). Washington also was not averse to playing Jordan against the UAR, e.g., to renegotiate the Gaza strip, "possibly turning it over to Jordan as a Mediterranean outlet with transit rights".[3]

By the end of the year, Washington had so chipped away at efforts to implement the Security Council Resolution that the talks—apart from serving as a channel for avoiding a big-power clash—were seen by many mainly as a screen for Washington strategy to divide and weaken the Arab states and to support some Israeli annexation of occupied territory —also aimed at separating Jordan and Syria from the UAR. Moreover, since mid-1969, when Israeli escalation on all fronts—artillery, missile, air and forcing the canal—had prompted U Thant to declare that, in fact, an actual war was going on, it was hard to escape the conclusion that US strategy was satisfied "to freeze" the situation at the existing level of hostilities which were advantageous to Tel Aviv.[4]

---

[1] US support of Israel, he stated, was creating serious internal problems in the Lebanon, Jordan and Saudi Arabia, Arab youth were becoming anti-American, while Soviet support was growing. A consequence especially of the US sale of Phantoms to Israel when various countries were beginning to place hopes on the Jarring mission. Interview in *Christian Science Monitor*, February 8, 1969.

[2] *The New York Times*, July 3, 1969.

[3] *The New York Times*, October 19, 1969.

[4] In the UN General Assembly, for example, Foreign Minister Mahmoud Riad took "a markedly more strident tone" than last year and Sudanese Prime Minister Babiker Nwadalia in his maiden speech attacked the United States with "virulence". Washington was accused of blocking a Mideast accord by supplying Phantoms and failing to insist on Israeli withdrawal from all Arab territory. *Herald Tribune*, September 24, 1969.

A US-backed Israeli offensive strategy of continuing armed hostilities, with the political and moral losses to fall mainly on Tel Aviv, and entailing no serious US efforts at political regulation within the Security Council Resolution framework, required, in the first place, Washington's maintaining Israeli military superiority, especially in the air.[1] Such a skewed military balance for a prolonged period was visibly unrealistic and almost bound to alter in favor of the Arab states in the long run. This was the judgment of many observers including, for example, Nahum Goldmann, President of the World Jewish Congress, in stating that "time was working for the Arabs".

Actually a significant change was registered during the third year after the war. The Arab reaction to Washington's supplying Tel Aviv before 1967 with the means for conducting a modern "electronic war", and then with 50 Phantoms and 100 Skyhawks in the 2 years after the war, had been to turn to the Soviet Union to protect Egyptian installations.[2]

As a result of Soviet aid, Secretary of State Rogers in spring 1970 declared: "a new factor has entered into the equation in the Middle East"[3]—a judgment borne out by the fact that after April 17, Israeli air attacks decreased and thereafter were restricted to the Canal zone. Penetration to targets—military and civilian—deep inside the country had been brought to an end by Soviet-provided anti-aircraft defenses.

Soviet military assistance to the UAR was paralleled by large-scale aid programs for economic development.[4] The defensive nature of such extensive support has been generally acknowledged even by opponents of Socialism. The weight of evidence, according to the Institute for Strategic

---

[1] "Without air superiority," notes a US military analyst, "it is doubtful that Israel could hold the east bank of the Canal with an accepted level of casualties." *Military Review*, January 1971.

[2] Interview with President Nasser, *U.S. News and World Report*, May 18, 1970.

[3] *The New York Times*, April 30, 1970; "U.S. Policy in the Middle East" by Bernard Reich, *Current History*, January 1971.

[4] Electrification for nearly 4,000 villages with power to be supplied by Aswan, a $700-million expansion of the Helwan iron and steel center, a $100-million phosphate complex at Qena, a smelter at Aswan, a ferrosilicon plant, and aid in drilling for oil in the Siwa Oasis. Czechoslovakia also was to supply and help install equipment for the construction in four years of the Kaft El-Dawar power station to generate 1.5 million kilowatts of thermal power, about half the capacity of Aswan.

Studies, argued "that the predominant Soviet concern was to reduce the danger of another locally unlimited war in the Middle East".[1]

The changed equation of force, referred to by Secretary Rogers in April 1970, was undoubtedly a major factor in accomplishing, on the one side, through Soviet military aid what Washington, on the other side, had been unwilling to do, by political and other means—exert pressure on Israel to halt raids.[2] Instead, the United States had been seeking to maintain Tel Aviv's superiority by trying to restrain the Soviet Union from providing arms to Arab forces in February and March 1970, according to President Nixon's policy statement.[3] That such attempts were rejected surprised no one familiar with the history of Socialist support for the national-liberation movement.

As a step toward political settlement in the newly developed situation, President Nasser took the initiative on May 1 in proposing to President Nixon either to influence Israel to withdraw troops or at least temporarily to halt US delivery of weapons. Secretary Rogers' counterproposal was for an Israeli troop withdrawal coupled with a 3-months ceasefire, during which Jarring presumably could seek a regulation of the crisis. The UAR, looking to an overall political settlement, accepted.

The resultant August 8 ceasefire took place, however, without Israeli troop withdrawals. Washington immediately seized the credit for the initiative, and apparently not unmindful of the opportunity of driving a wedge in the anti-imperialist Arab front as a result of this concession, did succeed in causing dissension on the Left—especially in Syria, Iraq and among the Palestinian guerrillas. In the following weeks, which witnessed a US threat to intervene in the Jordan crisis and Nasser's death, Washington exploited the reigning uncertainty in the Arab world by accusing the UAR and the Soviet Union of standstill violations. Under cover of

---

[1] *Strategic Survey 1970*, p. 9.

[2] On February 2, 1970, for example, after Abu Zaabal was subjected to bombing, Washington advised that if there would be no ceasefire, there would be "intensified Israeli raids in rear areas and America could do nothing about it". President Anwar Sadat, Speech at Helwan, May 1, 1971.

[3] "U.S. Foreign Policy for the 1970's," the *Department of State Bulletin*, March 22, 1971, p. 392.

this diplomatic and psychological offensive, Israel broke off talks with Jarring, and the United States sent additional ships to the Mediterranean in a demonstration of strength, agreed to supply more Phantoms to Israel and a new credit for weapons, and conducted reconnaissance flights over the UAR territory, which infringed upon sovereignty.[1] At the same time the "Rogers plan"—without seriously seeking to advance toward political regulation—could urge renewals of a ceasefire, which taken alone was tending to freeze a situation of low level armed hostilities and de facto occupation.

It was to break out of this limbo between war and political settlement that President Sadat proposed on February 4, 1971:[2] in return for a partial Israeli withdrawal to a line behind El Arish—the reopening of the Canal in six months, a prolongation of the ceasefire (to expire on March 7, and thereafter not renewed), guarantee of free passage in the Tiran Straits with an international force at Sharm el-Sheikh.

But, whereas the UAR proposal was explicitly made as a first step toward political regulation in the framework of the Security Council Resolution (which Tel Aviv basically opposed), Washington used the proposal as a point of departure but distorted its intention. The US plan treated the opening of the Canal in the form of an "interim agreement", but not linked with an overall settlement envisaging the complete withdrawal of Israeli forces. Such an agreement would, in effect, contribute to permanent Israeli occupation.

The follow-up Rogers' mission in the mantle of "honest broker" directly to such involved countries as Saudi Arabia, Jordan, the Lebanon, and the UAR and Israel in April-May 1971 sought to improve the US image in the Arab world[3] by lending support to an Arab ideological line of wooing or "neutralizing" the United States, which could both fan Rightist Arab nationalism and undermine the anti-imperialist front. Foreign Minister Mahmoud Riad, in an interview with *Le Monde* in mid-June, aptly characterized the trip as essentially a maneuver designed "to seduce" world opinion—with

---

[1] See Statement of Ministry of Foreign Affairs, USSR, *Pravda*, October 9, 1970.

[2] Interview in *Newsweek*, February 22, 1971; also *Herald Tribune*, February 15, 1971.

[3] In diplomatic preparation, Secretary Rogers, at his news conference on March 16, had stressed that Israel could not find security in geography, and called in general terms for a withdrawal to 1967 borders.

Secretary Rogers declining to exert pressure on Israel for a settlement and the United States continuing to supply Phantom fighter-bombers to Tel Aviv.[1] By braking Sadat's "friendship offensive", moreover, Washington also was contributing to provoking Arab forces concentrating on a military resolution of the conflict. This played its role in the ensuing Egyptian internal crisis, which broke out after President Sadat's return from Benghazi on April 17 and centered on the issue of Arab federation.

On an international plane Washington's defensive parries in the four-power consultations dovetailed with its contrasting aggressive maneuvers in the Afro-Arab states. By September 1970, a month following the UAR acceptance of a three-month ceasefire, US representatives, after pursuing dilatory tactics in meetings, emphasizing "quiet diplomacy" and evading substantive questions, finally broke up the "working group" by refusing to participate in the preparation of a memorandum for the UN Secretary General on the progress of the talks. This was paralleled by Washington's "persuading" Tel Aviv, which had withdrawn from the Jarring peace talks on the pretext of UAR and Soviet ceasefire violations, to return.[2] These talks, as could be expected, also recorded no progress towards a political settlement in the course of successive monthly ceasefire renewals, and Tel Aviv's return was characterized by President Sadat as a maneuver to avoid censure, and to draw out and perpetuate occupation for another 20 years.[3]

Washington provided further evidence in early February 1971 that it continued to back this general strategy when the US representative at the four-power consultations declined to support a proposal, agreed to by the other three powers, to underline the necessity for implementing the entire Security Council Resolution, particularly with respect to with-

---

[1] Thus, Washington, on the eve of the Secretary's departure, promised Israel 200 additional warplanes and $1 billion in credits for military and economic projects, according to *New York Post,* March 16, 1971.

[2] *The Wall Street Journal*, November 19, 1970. The instruments used, according to this organ, were fresh US arms shipments to Israel including 36 jet aircraft and some 200 tanks to "offset new Egypt missile sites built with Soviet help. Unstated, but clearly implied, is an American threat to slow arms aid to Israel".

[3] Quoted in *Daily World,* December 30, 1970.

drawal of forces from occupied territory.[1] The role of the United States in the Middle East, said President Nixon, was to "maintain the balance of power"[2] and it was not putting any pressure on Israel to make concessions.[3]

As long as Washington continued this fundamental positions-of-strength strategy, its participation in the talks could hardly serve to bring closer a political settlement and, therefore, perhaps, to avoid further political and moral losses to itself. For its key role in making Zionist expansionist policy realizable had become quite clear. "If the United States wishes peace," Nasser had declared, "it can get Israel to withdraw its forces from occupied territory."[4] And if Washington policymakers had nourished secret hopes that a successor to Nasser would change UAR policy, they were destined to be disappointed. For, President Sadat in no less uncertain terms indicated that the United States—if it so desired—could as in 1956 help "get Israel to withdraw from Sinai".[5]

In view of Arab awareness of a Washington strategy concentrating on the employment of military force—directly or indirectly—to turn the Arab and African states from their chosen path of development, US policy was risking shipwreck. Egypt, declared President Sadat, "waged a fierce struggle to repel the triple aggression in 1956, in effect, winning us the right to construct the Aswan Dam".[6] Such a gigantic step forward, achieved while armed hostilities were still going on in the aftermath of the second Israeli aggression, was convincing proof that the Egyptian people would not be swerved from a Socialist-oriented path which alone offers them the prospects of building a new life.[7]

---

[1] Since the summer of 1970, Washington had diluted its position on this question, and at the December 10 Press Conference, President Nixon evaded a reply on this question.

[2] Television broadcast, March 3, 1971.

[3] Ibid.

[4] Speech, May 1, 1970.

[5] Interview of February 10 with *Newsweek*, February 22, 1971.

[6] Speech at the opening ceremonies of the Aswan Dam, *Pravda*, January 15, 1971.

[7] In the UAR, for instance, the public sector "which, in effect, provides the economic basis for revolutionary-democratic policy" now embraces 85% of total industrial production. Nationalization of imperialist property has taken place in Algeria. Many foreign enterprises, banks, trading companies have been taken over by the government in Guinea, the Sudan, Somalia and Tanzania. Agrarian reforms have taken place

With time, therefore, it was becoming incumbent on Washington to show greater sophistication if it were to reduce—much less recoup—its own political losses in the Afro-Arab world. Thus, the State Department could be expected to voice more open criticism of Israeli inertia diplomatically since "in the absence of progress toward a political settlement there has been recurring evidence of arms shipments to the area".[1]

Politically, Washington's fears of growing anti-imperialist cohesion were somewhat allayed in the aftermath of the Rogers' mission by the windfall of a rash of Right Arab nationalism and anti-Communism in the Sudan and neighboring states. But, if there had been hopes of the Afro-Arab states splitting off from their Socialist allies, these were dimmed by the UAR-USSR 15-year Treaty of Friendship and Cooperation of July 4, 1971, and by the Arab and other voices raised to call a halt to the wanton murders in the Sudan following the reversal of the July 19 coup. (Washington, as well as Peking, to be sure, moved in with haste to establish more intimate relations with Khartoum.) Behind a "low-profile", the Nixon Administration did chalk up gains from the provision of Jordan with some $50 million of military aid—artillery, aviation and tanks, which enabled the monarchy to inflict heavy losses on the partisans in the latter part of July 1971—although as a by-product of this defeat, the leadership of the main guerrilla organizations rallied to form a single military command, financial fund and information unit.

With a changed equation of force and no renewal of the formal ceasefire, Washington had to convey a "sentiment of movement" toward political settlement by "defusing periodic outbursts", if the strategy of a "shooting peace" was not to erupt into a large-scale shooting war. Its diplomatic response to President Sadat's February 4 plan for reopening the Canal, for example, involved a document (later disclaimed, hence alluded to as the "Phantom Memorandum") given to Foreign Minister Riad by the chief American diplomat in Cairo, Donald Bergus, in which the United States appeared to endorse an Israeli pullback from at least half of the Sinai Peninsula to

---

in the UAR and Syria, have begun in the Sudan and Somalia, and are to begin this year in Algeria. In the Congo (Braz.) all land and its mineral wealth is state owned. L. I. Brezhnev, *Report of the Central Committee at the XXIV Congress of the CPSU,* March 30, 1971.

[1] *Herald Tribune,* April 13, 1971.

be occupied by Egyptian troops. But moves such as this—even if not disavowed—for a separate truce were interpreted as creating the legal basis for a de facto freeze of Israeli occupation elsewhere, and were rebuffed by President Sadat and later in the Soviet-Egyptian July 4, 1971, communiqué.

The Nixon Administration's continued commitment to force in the form of an Israeli military superiority was confirmed by such parallel moves as sending CIA chief Richard Helms in July to compare with Tel Aviv "the precise extent of Soviet and military power in the area", and was accompanied by hints of bringing Israel into a Western military pact. This could not help fostering the conviction that Washington planned no serious efforts towards an overall peaceful settlement, which, in turn, found reflection in President Sadat's repeated calls for the Arabs to prepare for a military decision in 1971.

With the issue of withdrawal of Israeli troops becoming politically more acute for the Arab Republic of Egypt on the central question of war or peace, certain differences between Tel Aviv and Washington positions surfaced. If Zionist leaders, such as Deputy Prime Minister Yigal Allon, were bellicosely avowing their expansionist aims,[1] Secretary Rogers, in his speech to the United Nations on October 4, 1971 on an "interim settlement", was diplomatically baiting a proposal for the opening of the Canal with a limited Israeli pullback. But without making a Suez agreement an integral part of overall regulation, the way was being left open for an indefinite Israeli occupation of other territories.

In the light of this tactic, Cairo felt it necessary in November to cut off negotiations with Washington representatives "because of their deceit and cheating and lies".[2] And, the following month, the impasse was further underscored when a four-nation African delegation to Israel and Egypt, led by Senegal's President Léopold Senghor, also made no headway in mediation. The US-Israeli blocking of a UN-envisaged political settlement, from all indications, was geared to fanning Arab nationalist leanings to military action or to foster-

---

[1] Thus, for example, "... we must become 'bi-national' simply by annexation. The Arabs must learn the lesson ... when they lose (the war, they cannot) get back all they have lost by political means." Interview in the *Nation,* May 31, 1971.

[2] President Sadat in his speech before the People's Assembly on May 14, 1972.

ing political moves to the Right and away from world Socialist support which could perhaps open the way to a deal with US imperialism.

* * *

If Egypt constitutes the major political and military force in the North Africa and Middle East geopolitical complex, there are two secondary groups of Afro-Arab countries which play an important role in US imperialist strategic thinking: in North East Africa—Libya, the Sudan and Somalia; and in the Maghreb (West)—Morocco, Tunisia and Algeria. These complement the countries lying opposite—from Portugal and Spain in southern Europe to Turkey—tied to Washington either by separate agreement or NATO, and forming, in its view, a distinct regional complex around the Eastern and Western Mediterranean communications seaway linked by an 80-mile passageway between Sicily, Malta and Tunisia.

Politically, the British influence in North Africa, which was paramount in the East and bound up with oil exploitation and the strategic Suez passage to the Red Sea and Indian Ocean, and the French interests, which predominated in the Maghreb, were forced to give way to independent states in the postwar period, and to a US imperialism which sought to incorporate both its partners and their former empires in its sway. However, such a grandiose design was destined to last little more than a decade after the Suez triple aggression before it, too, became seriously undermined in the face of the growing national-liberation movement, sharpened anti-imperialist struggle, and mounting influence of Socialist ideas.

These three generally mutually supporting dynamic forces, especially in the polarization following the June war, helped to generate the coup in Libya on September 1, 1969 which ousted King Idris, who had been placed on the throne in December 1951 with British and US support. Their main artery of control had been via a military presence and bases: Wheelus Field, the biggest US foreign base, built under a 1954 treaty, with a complement of 3,000 men and used for the bi-annual training, gunnery and bombing practise of some 21 air squadrons from European bases (West Germany, Britain, Italy, Spain and Turkey); Britain maintained an air staging base in El Adem and an armor and infantry "desert warfare" training base at Tobruk. There also was a special

7,000-men British-trained security force for the king (established apart from the Libyan Army), and a secret agreement had been signed by Britain on July 27, 1964, according to *Al Ahram*, which provided for land, sea and air intervention to protect the king "against internal troubles and external dangers", to be carried out together with US troops.

Following the successful coup, which caught the monarchy by surprise and foiled imperialist intervention, nationalist sentiment was immediately directed to halting foreign training flights and then forcing the evacuation of bases in early 1970. With the foreign mailed fist removed, Libya proceeded to harness US and British monopolies' unbridled control of the country's enormous oil wealth and thereby to gain an increased share of the profits from crude oil—nationalizing marketing and distributing interests of Esso (US), Shell (Anglo-Dutch) and Anseil (Italian) on July 4, 1970 and foreign banks by the end of the year. In an effort to prevent this contagious anti-imperialist movement from further spreading, Secretary Rogers announced on June 26 that he hoped "to persuade the British to retain troops in the Persian Gulf after 1971", thereby confirming once more the affinity of military bases for oil.

Whereas the US and British military presence in Libya had nakedly represented the mailed fist of imperialism, its influence was more cloaked in the Sudan, to which Washington attached strategic importance as the southern flank neighbor of Egypt, a threat to imperialist influence in the Red Sea area, and a vital link between Arab and Black Africa. The significance of this link was shown, for example, in the events following the US-Belgian-British intervention in the Congo in November 1964 when the Congo's lines of communications with friendly African states were effectively disrupted.

This was not unrelated to Washington's and London's persistent attempts to prevent the emergence of a politically unified Sudan, e.g., in the rebellion going on since 1955 waged by southerners (mainly Black, who are Christians or animists) against northerners (mainly Arab and Moslem). Although ethnic and tribal tensions in the Sudan go back for centuries, they had been constantly fanned under colonialism[1] and were still being played upon by the imperialist pow-

[1] British policy since the nationalist upsurge in Egypt in 1919 was to treat the Sudan's South separately from the North with the possibility of integrating the South with East Africa as a counter to a united Nile

ers.[1] The possibility of a socio-ethnic solution to the country's divisive internal strife existed briefly after the overthrow of the military government of General Ibrahim Abboud in October 1964, but was never realized following a reactionary offensive which led to the expulsion of Left-wing forces from the government in 1965. By the end of 1966, the 3 southern provinces were seeking—reportedly with CIA financial support—to form a separate state as the Republic of Azania, which American diplomats were urging African states to recognize.[2]

In an effort to steer clear of dependence on imperialism —mainly since the June war—the Sudan made various aid and trade agreements[3] with Czechoslovakia, Algeria and the Soviet Union, as well as with Saudi Arabia, Italy and the IMF. Politically and militarily, the Sudan turned increasingly against the Washington-Tel Aviv alliance and more to the Arab world and Socialist states for its own support. Thus, realizing that the United States was Israel's major patron, the Sudan reacted sharply to the aggression by breaking off diplomatic relations with the United States and at the conference of heads of states and governments of Arab countries in Khartoum came out strongly for joint action on the part of the Arab countries and later sent a military contingent to the Canal Zone. Military equipment and aid came from the Soviet Union and pilots from the UAR in 1970. Trade with the Soviet Union steadily increased and by 1970 was double the level of 1967. On the other hand, despite the break in Sudan-US relations, the International Monetary Fund "reluctantly" extended additional credits and then concluded a new $24 million loan in 1968 for increasing electric power supply. Khartoum, nevertheless, rejected the Bank's recommendation to split up the land on the Gezira Agricultural Scheme into private tenancies, which would have dissipated public control.

Within the compass of intensified Arab nationalism fol-

(Egypt and the Sudan). See, for example, Keith Kyle, "Sudan Today" in *African Affairs*, Journal of the Royal African Society, July 1966.

[1] Thus, the Sudanese Minister of Interior claimed that his government was in possession of documents substantiating the charge that the Western powers materially and morally had supported the troubles in the South. (*North Africa*, September/October 1967).

[2] *Za Rubezhom*, December 23, 1966, quoted from *West African Pilot*, Lagos.

[3] Details in *Africa Report*, January 1968.

lowing the coup on May 25, 1969, a Right-wing trend in the Revolutionary Command Council was early manifested in the banning of the Communist Party. This led to a weakening of the anti-imperialist front of workers, advanced peasantry, students and revolutionary intellectuals. It brought in its train, at the same time, an activization of semi-feudal and conservative bourgeois forces, such as the Umma party of the reactionary Imam el-Hadi el-Mahdi and its main support, the Moslem Brotherhood. In March 1970, according to Gaafar el Nimeiry in an interview in *Al Ahram*, the United States and other imperialist powers were involved in fomenting unsuccessful uprisings of Mahdists in a number of Sudanese cities, as a result of which the American commercial attaché (working in the Netherlands Embassy in the absence of US-Sudan diplomatic relations) was expelled.

The growth of Right-wing nationalist tendencies and activization of reactionary circles in the Sudan in 1970 and 1971 found encouragement on the part of imperialism and Arab reaction abroad to the detriment of the country's progressive development. This led to intensified inner-political struggle. Under the influence of Rightist circles in the Sudan, an offensive was launched against progressive forces which led to the tragic events of July 1971 and weakening of the country's anti-imperialist effort.

Eager to exploit this wedge further, Washington pledged to the Sudan $27 million in 1972 "to aid refugees from the civil war", the Export-Import Bank—$3.3 million for road-building equipment (and was considering financing the purchase of five Boeing jets); and the IMF advanced $40 million in credits. On July 20, Khartoum resumed diplomatic relations with the United States.

The 1969 revolts in the Sudan and Libya also had exercised a catalytic effect in Somalia in overthrowing on October 21, 1969 the pro-Western government of Prime Minister Egal (who had swerved away from the neutralist course of President Shermark, assassinated earlier in the year), only a few days after his return from a visit to Washington. In the Arab nationalist struggle to free the region from the imperialist grip, the Sudan had been providing technical, cultural and military support for Somalia, an object of US and British strategic interests to control the Gulf of Aden. The struggle of Somalia for nationhood in the mid-sixties also had involved fighting to unify with its three-million population, the Somali nomadic tribesmen, now living mostly in Eritrea (under Ethiopian administration since 1962, and where the US base at Asmara is located) and also in Kenya.

In view of the strong influence of the United States in neighboring Ethiopia and the British in Kenya, the Somali Republic had felt its independence best served by rejecting, at first, a US offer of arms and planes and turning to the Soviet Union for aid. (This was not unaccompanied by the usual Western clamor of Soviet penetration.) By 1966, Somalia was receiving aid from various countries: Italy had provided aid estimated at £45 million; Italy and the United States had equipped the small, well-armed and relatively independent Somali police force, which was trained in and had ties with the United States and the Federal Republic of Germany; and US aid was only slightly less than that of the Soviet Union. In November 1967, peace with Kenya, as the result of the Kinshasa OAU meeting in September 1967, was signed in Addis Ababa eliminating a major obstacle to further Western influence, and was followed by Vice President Humphrey's out-of-the-way visit to that small country in January 1968 and the signing of an $8.5 million aid agreement. This disproportionately large program suspiciously was accompanied by government restraint on popular anti-American imperialist feeling.

But dollar aid with political strings proved insufficient to counteract the more dynamic nationalistic and anti-imperialist forces exacerbated by the "no war, no peace" strategy of Washington and Tel Aviv, leading up to the 1969 military coup. The new Left-wing government proclaimed adherence to scientific socialism, the "Somalization" of financial and trade organizations, and recognized the German Democratic Republic and the progressive governments of South Vietnam, Cambodia and North Korea.

In neighboring Eritrea, too, the aftermath of the June war found Afro-Arab countries providing aid to the Eritrean Liberation Forces. (Moslems comprised some three-fourths of the three-million population of Eritrea, turbulent since annexed by Ethiopia). Thus, not surprisingly, the parallel US-Israeli support to opposing forces "seemed during 1970 to be becoming almost an extension of the Arab-Israeli war".[1]

US postwar political-strategic aims in the Maghreb, conditioned largely by the prolonged and bloody Algerian li-

[1] *Strategic Survey 1970*, pp. 51-52.

beration struggle, focused on the two flanking countries—Morocco and Tunisia. But, whereas American imperialism was generally allied with the British in their former sphere—from Libya eastward, it was more competitive if not hostile to the French in the Arab West. Moreover, since it was not engaged as the immediate colonial power, it became less embroiled than France, which, for example, under pressure of the Tunisian armed forces to evacuate the Bizerte base, replied with a bloody attack on July 20, 1961 but was soon forced to withdraw with great political and moral loss. Washington's more sophisticated tactics were to agree to the evacuation of its bases in principle, but to draw out negotiations (e.g., as in Libya for almost a decade), or by other means to circumvent their intended effects.

Thus, in Morocco, although the United States formally announced the closing down of its $400-million strategic air force base at the end of 1963 (in accordance with the 1959 agreement between President Eisenhower and King Mohammed V), it called attention openly to its overflight rights but kept secret for seven years a "private arrangement" to retain a large naval communications center at Sidi Yahia, 50 miles northeast of Rabat, which duplicated the Pentagon-constructed facilities oppositely at Rota in Spain. The United States also continued its covert naval base with some 1,700 men to service the Sixth Fleet at Kenitra (20 miles from Sidi Yahia), according to information released by the Senate Foreign Relations Committee in mid-1970, who were publicly referred to after 1963 as a "training mission". This was part of a cover story arrived at by the two governments to conceal the existence of an American base, which the State Department subsequently periodically and categorically denied.

Similarly, in Tunisia, the United States moved in after France and established a new US base near Bizerte for servicing the Sixth Fleet. The agreement was signed in 1966 by an American group representing Tampa Ship Repair to build a naval repair yard out of the former base and arsenal at Menzel Bourguiba, to be run by a mixed US-Tunisian Co. SOCOMINA with a $10 million investment—ostensibly for foreign oil and other cargo vessels.

Closely meshed with Washington's prime strategic interest has been its economic and financial penetration through aid, surplus food and international credits (earlier discussed), which have led to growing indebtedness and dependence. Politically, both Tunisia and Morocco are conspicuous in withholding criticism of US aggression in Vietnam, with President Bourguiba openly supporting Washington and buffering Tel Aviv, calling for the resignation of President Nasser in October 1967, and seeking to turn the OAU into an advisory body and liquidate its Liberation Committee. Washington's close ideological ties to leaders such as Bourguiba extend to personal friendship and the provision of private medical attention.

In post-liberation Algeria, Washington, no more than Paris, could hope to re-establish a foreign military presence. At best, initially, it could hope only to neutralize and modify the foreign policy of this central country in the Arab West. And this, in fact, was the aim of US diplomacy in applying crude economic pressures, e.g., withholding food aid and signing small agreements up to early 1967. Such attempts to swerve Algeria from its anti-imperialist course did not avail, and in the aftermath of the Israeli blitzkrieg, Algiers continued its militant policies, including a step-by-step nationalization of oil, the basis for which had been laid through years of critical Soviet aid in providing the equipment, technology and training which imperialism was loathe to supply.

Such new-found economic and technical strength of the Afro-Arab states, which in no small part also was a corollary of the overall military, political and diplomatic support provided by the Socialist world, enabled them for the first time to launch a concentrated offensive against the monopolies, to raise the prices and to gain control of their most valuable natural resource. In two major bargaining battles in early 1971, first the Persian Gulf states at Teheran, and then Libya at Tripoli forced a greater sharing of profits amounting to $15 billion for a five-year period.

Libya, which had become Europe's leading petroleum source after the June war, was demanding in January 1971 a 68% rise over the 30 cents per barrel increase to $2.53 it had won in mid-1970 from Occidental Petroleum following a cutback of output and crucial squeeze on Europe's supply of oil. (The Idris government, even before September 1969, had sought a modest 10 cents/barrel rise, but the US and British monopolies were still only offering an increase of 6-10 cents/barrel in May 1970.) At Tripoli, Libya was able to obtain an increase in crude from $2.55 to $3.45 (at the Texas wellhead, a barrel sells for $3.40), or more than double the increase won at Teheran, and thereby to lift its revenues from $1.3 billion in 1970 to $2 billion in 1971. Curiously, Libya held bank deposits in Washington exceeding American capital investments in Libya.

In nearby Algeria, Sonatrach, one of the largest state oil enterprises, found, as a "counterweight" to French oil companies, new clients such as Mobil and Shell in the United States with which it contracted at the close of 1970 to supply 60 million tons of crude over a four-year period. Following nationalization in February 1971 and a subsequent six-month French boycott, US companies stepped into the breach to conclude the world's biggest deal to supply oil and oil products to Commonwealth Refining Co. for $8 billion over 25 years. This, together with deliveries to El Paso of liquefied natural gas, would help to ensure the insatiable needs of US industry over a long term, to tie Algeria to American eastern coast markets, and perhaps to exercise a political influence aimed at since the days of the Kennedy Administration. This was paralleled, not surprisingly, by Department of State approval for a $250-million loan for a gas liquefication plant and an unprecedented loan for the purchase of US jet airliners. The chain reaction of nationalization swept back to Libya in December 1971 when it took over British Petroleum in which the British government has a near majority interest. As a result of the Soviet-Libyan agreement of March 1972, providing for joint development and refining, oil was first loaded on a Soviet tanker on June 2, thereby breaking the boycott imposed by Western oil companies. The wave of nationalization also overtook the Iraq Petroleum Co. when Baghdad became the third country within a twelve-month period to gain control of its natural wealth on June 1, and gained immediate support from the other Arab states and Socialist countries.

Such unprecedented actions in the Afro-Arab world indicated that the imperialist "no peace, no war" strategy was far from bringing its proponents only desirable results, particularly after the changed military balance in the spring of 1970.

In glancing briefly at major elements of Washington (and usually Tel Aviv) strategy since the changed balance in early 1970, one is struck by certain features appearing repetitively in both official Statements and actions. Politically, the determination above all to break away the Afro-Arab states from close cooperation with the Socialist countries—by placing the Soviet Union as a "big power" on the same footing with imperialist states, or portraying her as a successor of Russian Tsarism; by fanning Right-wing nationalism against anti-imperialist internationalist forces (the natural and major allies of the national-liberation movement);[1] by inviting a deal with the United States, since it holds the "key" to settlement.

Secondly, militarily to continue and even increase support for Israel, a dependent and dependable ally, as a cornerstone of regional force against progressive states. To bring US armed strength to bear at various critical political and geographical points, such as behind Jordan against the Palestinian guerrillas and Syria in 1970. To expand its naval facilities in the Mediterranean, e.g., acquisition of Piraeus as a base in February 1972 after being forced to evacuate Wheelus Field, exert pressure on Cyprus for NATO bases, and to help conclude a new and broadened Malta agreement on March 26, 1972.[2]

Finally, to take advantage of the generally acknowledged Soviet desire for peace and to avoid military confrontation—in the common interests of the Socialist community, national-liberation movement and popular masses in the capitalist world—with the premise "endorsed by successive American Administrations that if the Kremlin found unassailable barriers in its path, it would accept these philosophically and accommodate itself to them".[3] If such was the underlying premise of Washington policy in the sixties, it is even more dangerous and clearly inappropriate to the situation in the seventies.

---

[1] On this vital question, L. I. Brezhnev stated unequivocally: "The entire course of events has shown that friendship with the Soviet Union provides the necessary support and assistance to the progressive Arab states in their most trying times. This is well understood in Egypt and in Syria, and in Iraq, and in Yemen. We have a treaty of friendship with Egypt and Iraq and we shall develop our relations with these countries on the basis of these documents. We are fully determined to strengthen our friendly ties with Syria, Algeria and other Arab countries as well." The Joint Session of the Central Committee of the CPSU, of the Supreme Soviet of the USSR, and of the Supreme Soviet of the RSFSR, December 21, 1972. *Pravda,* December 22, 1972.

[2] The British agreed to pay an increased annual rent of £5.25 million; but further annual payments of £8.75 million plus somewhat more in aid was to come from NATO countries. *Strategic Survey 1971,* London, 1972.

[3] Aaron S. Klieman, *Soviet Russia and the Middle East,* The Johns Hopkins Press, 1970, p. 98.

IV. CONCLUSIONS AND PERSPECTIVES

To interpret and assess the relationship of the United States to Africa is in essence an attempt to measure the application of a part of its global strength to a continent viewed as part of an organic world. Since the ratio of foreign power to African strength is so disparate, the continent has been and remains exceedingly vulnerable, and thus the critical influence of US imperialism in its efforts to "sway the balance", especially since Suez.

The apportionment of US strength, however, involves not only global and regional power and balance changes, but also considerations of the proportions and form in which force can be brought to bear with respect to the stakes involved. This presupposes an integrated evaluation of all spheres including highlights of: the primary political aims and policies of US ruling circles, the slower moving economic monopoly interests, the social strata affected, the ideologies based on these and world influences, as well as some important military aspects.[1] The interrelationship, it may be noted, is also reflected in microcosm as a "nesting" within spheres. Thus, for example, the category US "aid" reveals itself as a multi-dimensional model—a political-economic composite emphasizing military aims and showing a trend from a tainted bilateral to a more homogenized multilateral form to achieve greater effect per dollar.

---

[1] Cf. what may be regarded as a one-sided emphasis, "Neocolonialism: colonialism operating entirely through economic relations, instead of as before through economic relations accompanied by political domination." Pierre Jalée "The Third World in World Economy" in *Monthly Review,* March 1971.

Regionally, the priority and political-military emphasis given by Washington to the Afro-Arab states flowed from its early postwar strategy "against Communism"—in effect, the anti-imperialist forces—in the Mediterranean region, expanding in scope from southern Europe and Turkey and then increasing in intensity along with American monopolies' mushrooming profits in the Middle East. In tropical Africa, too, it was the political threat to imperialism as a whole posed by a successful Congolese national-liberation movement led by Lumumba which enticed Washington to exploit its global position in the United Nations to vie for a dominant influence in the Congo—first confronting its rivals militarily and then reaching a modus vivendi for joint exploitation of the country's enormous mineral wealth. In southern Africa, by and large political-military considerations also underlay US critical support for Portugal, partially screened via NATO, and Washington's (behind London's) buffering of South Africa as the strongest imperialist ally on the continent paralleled closely US monopolies' (also second to British) profitable economic ties. Closely related to these priorities and emphases, it may be noted, there are several conflicting currents among American bourgeois ideologists and policymakers which must be touched on—without attempting a broader socio-political analysis which would take us too far afield.

US policy toward Africa, which has been made rather pragmatically by American political leaders leaning on advisers who frequently also have verbalized or justified their courses in "theoretical" form, derives mainly from their foreign policy conceptions of strength. Without going beyond the framework of imperialist politics and capitalist economics, these conceptions differ qualitatively and may be categorized even if somewhat arbitrarily, since there is overlapping, into official views, "moderate" political criticism, and bourgeois reformism.

Official views of leaders of both Democratic and Republican Administrations, representing dominant monopoly interests in the military-industrial complex and imbued with the arrogance which comes with power, are proponent in fact, if not always in words, of political-military force. The military side, moreover, is frequently overemphasized, sometimes to the point of being exalted above other spheres. With variations depending on the period (discussed later) such views have been voiced in the past decade among others by

Rostow (earlier quoted) serving under two Administrations. Former President Johnson, whose training and Congressional preoccupation with naval and military appropriation biased him towards a "military view of world events", was a rigid advocate of "hard-headed" methods in selecting the "right levers for bringing US influence to bear".[1] Under a statistical charisma frequently referred to as "realism", George Ball, former Under Secretary of State, has defined a world power today as "only a cohesive society with a population approaching 200 million and a national income of $300 billion",[2] and also enjoying "arsenals of nuclear weapons, which do not—as some have foolishly argued—tend to equalize the power of great and small nations, but on the contrary, have quite the opposite effect."[3]

Although Robert McNamara was for a long time Secretary of Defense, than whom none should be a stronger proponent of the political-military emphasis, he was, nevertheless, acutely conscious of the limitations of military force, the limits of the US budget, and the key role played by economics in the developing countries. (Perhaps not unrelated to his New Frontiers' experience and the failure of the war in Vietnam was his later resignation to become President of the World Bank, seeking to convert development financing into political and economic advantage.) In the grouping here presented, McNamara would represent a transition to the second category: political criticism of official views.

The "moderate" critics of the official emphasis on political-military force received their greatest impulse, perhaps, from the failures in Vietnam in the early 60's, for which Washington's escalated answer, nevertheless, was to go further over to a "broad commitment of giving priority to the military aspects of the war over political reforms".[4] This critical group, sometimes misnamed "neo-isolationists", would substitute to various degrees a political-economic emphasis—political tutelage of the developing countries from traditional societies to capitalism and economic aid, with greater emphasis on peaceful construction. In Congress, members like Fulbright and Church would fall into this category; in the information

---

[1] See, for example, P. L. Geyelin, *L. B. Johnson and the World*, London, 1966, pp. 31, 269.

[2] George Ball, *The Discipline of Power*, p. 17.

[3] Loc. cit., p. 14.

[4] "Pentagon Papers," *The New York Times*, July 6, 1971.

media—Lippmann, writers in *The New York Times* and *Washington Post*. There is a mild recognition of the necessity for socio-economic change to achieve modernization, but more important is their vigorous advocacy of trading on US economic strength for political advantage, and also encouraging Rightist and nationalist trends in the Socialist world and developing countries.

The "Liberals", or bourgeois-reformists, who would achieve US foreign policy objectives by placing more weight on socio-economic reforms, and making greater concessions to the developing countries in competition with Socialism, are the least influential of the three groups. Representatives of this category have included some of the "new Africa" members of the Kennedy Administration, such as Chester Bowles, Africanists in universities and foundations, and certain Afro-American groups. They have had, however, only a marginal influence on governmental policy—greatest in the early 60's—and were most useful in presenting America's best face to Africa at the high tide of its successes.

It is of no little significance that these three groups concentrate on different spheres of strength and also recognize that the totality—without being an arithmetic sum or simple equation—is most vital. Moreover, their differences—from the position of the "Rightists", on political-military force oriented almost exclusively to foreign and domestic oppressors, to the position of the "Leftists", on the socio-economic plane and showing a greater degree of accommodation to rising national and class forces—are taken into account by both Americans and Africans. Although their common and fundamental weakness derives from the systemic inability of US imperialism to renounce its aims, it would be rash to say that it cannot downgrade its objectives or change its sphere of emphasis, as has been demonstrated by the experience of the past decade.

How much force Washington has been willing or able to bring to bear against anti-imperialist African forces has been roughly geared to the ebb and flow of the political-military tide on the continent and globally. In the period of ascendancy of world Socialist and national-liberation forces and changing world balance, coinciding in Africa with the British and French imperialist defeats in Suez and Algeria, Washington entered the breach and found it expedient to place greater emphasis on economic levers. If these proved insufficient to seduce or cajole popular movements, especially in the Congo—the watershed between independent and subjugated Africa—Washington, relying on a somewhat enhanced "power-image" after the Cuban crisis, appeared less hesitant to apply force.

The US-led joint intervention at Stanleyville in November 1964 may be seen as a marker in this swing of the compass needle in the direction of military force. This was followed by a rash of reactionary coups (and a more sympathetic Washington attitude toward settler Africa), which left as the major progressive force on the continent the Afro-Arab states. The Israeli-Arab war, an aggression of which President Johnson "knew beforehand" and "personally approved", according to President Sadat,[1] represented, for all its complexity, a continuation of this trend. The limited political nature of this military victory, however, came to light in the war's aftermath. A critique of US policy in the Afro-Arab region, the primary concentration of Washington in Africa, becomes at the same time a critique of the official emphasis on political-military force. The present strategy, as has been indicated, is leading to greater radicalization—despite ups and downs—of Arab forces, polarization against US-Israeli imperialism, and the possibility of a new and wider round of war. Such a war could scarcely end more favorably militarily for Tel Aviv or Washington, and probably less so politically and economically—to say nothing of its disastrous consequences for the peoples involved.

The increasing difficulty of maintaining an equilibrium of "no peace, no war" between two dynamic nationalisms—a state which was viewed by U Thant in autumn 1969 as looking like the beginning of "the Hundred Years' War", led him to change two years later to the judgment that if the present impasse persisted, new fighting would break out "sooner or later" to a full-scale war, with the danger of its spreading and involving other powers.[2] Although this pointed to Tel Aviv's adamant refusal to change its position "on the question of withdrawal",[3] as the immediate cause it was clear to many which power was making the Israeli position possible, as well as the logical way to break the deadlock.

[1] Speech on radio and television, Cairo, September 18, 1971.

[2] Introductory remarks to *Annual Report* of the 26th UN General Assembly, September 19, 1971.

[3] Ibid.

The Security Council Resolution 242, which took account of the issues of principle (i.e., the rights of the peoples) and the balance of forces—both regionally and internationally, continues to provide the basis for regulation—even if expansionist Zionism cannot be expected to give up voluntarily its policy of occupation of territory or expulsion of peoples, a policy which it has, by itself, neither the population nor strength to enforce. "A settlement will one day be reached," wrote Lord Caradon, a negotiator of the 1967 Security Council Resolution, "the question (is) whether that settlement can be reached in peace or whether it will be reached only after terrible bloodshed."[1]

If a political settlement is to be reached, it can scarcely be expected to come about as a spontaneous process. "Many states," declared L. I. Brezhnev, "have spoken out for a solution of Middle East problems on the basis of the well known UN Security Council Resolution. Unfortunately, however, verbal support alone does not suffice. If it were underpinned with concrete political actions, Israel would be forced to accept peaceful regulation, to recognize the lawful rights of the Arab peoples. As for the Soviet Union, our readiness to do our part is well known."[2]

Washington's strategy since the Congo intervention, even after correcting a number of previous mistakes, has several major flaws. Although avoiding the past political disasters of direct imperialist aggression (e. g., Suez, the Congo, Vietnam), US policymakers have nevertheless, underestimated the role of the Arab national-liberation movement as an independent, anti-imperialist political force, as well as its ability to see through the US-Israeli direct support relationship. US official views also have overestimated the limits of military force and particularly the value of military surprise, or blitzkrieg, with its short-term advantage of winning battles but lacking the long-term capacity to win the war.[3] The

---

[1] *War/Peace Report*, December 1970.

[2] The Joint Session of the Central Committee of the CPSU, of the Supreme Soviet of the USSR, and the Supreme Soviet of the RSFSR, December 21, 1972. *Pravda*, December 22, 1972.

[3] The strictly military argument for a lighting-type war at Suez, given by Prime Minister Eden to President Eisenhower, for example, was that "only swift Israeli military successes and Anglo-French action within 24, or at the most 48, hours could prevent the spread of the war to Syria, Jordan and Iraq." Eisenhower, op. cit., p. 76.

primacy of the political, on the other hand, has been demonstrated convincingly more than once—in Vietnam and the Afro-Arab world in recent years, not to go back to World War II for a momentous historical parallel (when, incidentally, the role and all-round strength of the Soviet Union was also badly underestimated). Furthermore, the possibility of an expanded war, which President Eisenhower was quick to appreciate when other imperialist powers resorted to, or were the patrons of, naked military aggression,[1] apparently is not being given due weight today although no less real.

When the Nixon Administration, deferring to political losses incurred from its military overemphasis, turned greater attention to the political and diplomatic fields in early 1970, it quickly proved to be for purposes of maneuver rather than to alter its strategy or conception of strength. By dissociating from Zionist occupation and refugee policy, Washington (like London) might hope to escape paying a slowly mounting political price, but without decreasing support for Tel Aviv it was not lowering the pitch. By supplying new arms, including M-60 tanks to Jordan in mid-1971, more as a military threat to Syria than for use against Palestine commandos, according to President Sadat, Washington was actually increasing the dangers of a new outbreak. By fanning defeatism, Right-wing nationalism and "anti-Communism", the Administration could record some weakening of Afro-Arab national and international unity—with calls heard to balance policy between the imperialist and Socialist powers, but at the same time to concentrate all attention on Israel for a military solution.

In contrast to the prolonged armed hostilities in North Africa, the seductively successful US-led military intervention in the Congo in November 1964 was followed by a political stability favorable to Washington, whose efforts then turned largely to political and economic matters. Here, like in most of the relatively weak states of tropical Africa, where tribalism and traditional society are widely prevalent, the long-

---

[1] After the Anglo-French ultimatum to Egypt on October 30, 1956, for example, Secretary Dulles lashed the French Ambassador with the angry recrimination: "This attack on Egypt incurs the risk of a general war." See Herman Finer, op. cit., pp. 6 and following.

term struggle for genuine independence bound up with economic development and nation-building is especially hindered by neocolonial blocs and ties. Hence, the great significance of political and economic steps to achieve greater African strength through unity, such as the formation of the OAU in 1963 and regionalism since 1967,[1]—neither of which, however, is free of US and other imperialist influence. Nevertheless, Washington cannot have failed to notice the tendency of black Africa gradually to close ranks with the Arab countries since the June war,[2] to view the continent increasingly as a whole, and to show greater appreciation of the outcome of the armed struggles to the north and to the south against a common enemy.

This is particularly felt in southern Africa, where the armed struggle for independence from Portuguese colonialism since the early 60's and partisan warfare against racism in Rhodesia and South Africa since August 1967, have compelled US imperialism, which is politically and morally insecure as a major supporter of oppressive minorities, to make determined efforts "to rehabilitate" the apartheid régime politically in the eyes of black Africa and the world.

A partnership which closes the triangle with Washington is Israel and South Africa, both in regions of largest US investment in Africa, both intruder minorities holding down by force of arms large exploited majorities. Prime Minister Vorster has drawn the analogy that "Israel is now faced with an apartheid problem—how to handle its Arab inhabitants" which he viewed "with understanding and sympathy".[3] This solidarity, moreover, extends to a military understanding whereby, for example, South Africa manufactures the Uzi submachine gun—an Israeli invention with license from Belgium, and Israeli blueprints of Mirage fighter engines were reportedly made available to South Africa. Praetoria and Tel Aviv both feel that they are outposts of the West.

---

[1] Ya. Etinger, *Political Problems of Intergovernmental Relations in Africa*, M., 1970; the author periodizes on this basis: 1958-62—prior to the OAU; 1963-66—the OAU; and since 1967—regionalism.

[2] The OAU session in Addis Ababa in June 1971 urged African states to take practical measures to compel Israel to withdraw from occupied territories. This, despite Foreign Minister Eban's declaration that the "Middle East crisis is not an African problem, and Africans should not become involved in it".

[3] *The New York Times*, April 29, 1971.

* * *

In sum, from the standpoint of US strength and global policy, what might one be led to expect of the US relationship to Africa in the 70's, a period when US world commitments are generally recognized as being overextended.

This overinvolvement has been voiced, perhaps, most clearly in the US Doctrine, elaborated at Guam in the summer of 1969 and then in President's Address to the Nation of November 3, 1969. In redefining Washington's role, he pointed to the "growing strength and autonomy" in other countries, and domestically to the "nascent isolationism in reaction to overextension".[1] To continue present US policy, he noted, "certainly would have exceeded our psychological resources", even if, as he equivocated, it "might not have been beyond" US physical resources. Hence, an increased emphasis would be placed on allies, who must "assume the primary responsibility of providing the manpower".[2] The United States will act "as *a* weight—not *the* weight—in the scale" (original emphasis).[3] Thus, with respect to direct American manpower participation, amount of financial involvement, and image at home and abroad, the United States would seek in future to project a low profile.

Although the new formulation of Administration strategy may have been precipitated by the eroding failures of US armed forces as a substitute for an inadequate Saigon ally in Vietnam, it was already being applied more "successfully" with respect to Washington's ally, Tel Aviv, against Afro-Arab states. Here, too, the rather grim essence of this up-dated alliance policy was for the United States to provide the military and economic backdrop and assistance and thereby help to avoid the political losses and moral obloquy which then would fall to a smaller expansionist partner.

In Subsaharan Africa, where the weaker independent tropical African states are caught in the bind of striving for economic development but lacking in capital, technology and skills, Washington's main sphere of pressure has been gen-

---

[1] *U. S. Foreign Policy for the 70's*, Report of President Nixon to Congress, February 25, 1971.

[2] Ibid.

[3] Ibid.

erally economic, for example, by reducing concessionary bilateral aid, such as it is, and by promoting instead more rapacious private investment. Although the latter has been growing in recent years at an annual rate of 12-14% "to encourage efficient development of Africa's resources of petroleum, mineral and agricultural products",[1] it has, as generally acknowledged, generated little industrialization, qualified specialists, or a balanced economy. Furthermore, if the President in February 1971 had promised tariff preferences "to open up new markets",[2] the US currency revaluation in August 1971 had, on the contrary, added substantially to the trade difficulties of African and other developing countries. In three important countries (Ethiopia, Zaïre and Ghana), US military aid apparently was the major channel of achieving Washington's aims. These economic and military levers were geared, on the whole, to political concessions in the individually vulnerable tropical states, which are too concerned, in the language of the presidential address, with "a jealous protection of their absolute sovereignty".

In southern Africa, Washington's "alliance policy" in the 70's forebodes no radical departure from its entire postwar links with both its NATO colonial ally, Portugal, or its economic partner, South Africa. To the former, which receives most of her arms (used in African colonies) through NATO but cannot finance her draining colonial wars, Washington is bent on granting large credits in the guise of payments for bases. The pivotal reactionary force by far, however, is the racist regime in South Africa, which received during the sixties about four-fifths of the US private capital going into the continent's manufacturing industry, as well as heavy machinery and "knowhow" through mushrooming trade, thereby adding to its industrial, technological and military advantage (even if the United States itself is not a significant Western direct arms supplier) over self-governing Black Africa.

Economic partnership, moreover, is being reinforced under the political umbrella of the Administration's covert policy adopted in early 1970 of expanding contacts, or a "dialogue". Political rehabilitation, indeed, would be a step in

---

[1] *U. S. Foreign Policy for the 70's*, op. cit., p. 104.
[2] Op cit., p. 117.

line with making South Africa a regional sergeant-major and, more far-reaching possibly, a connecting link between NATO and SEATO. President Nixon, apparently, "has accepted the strategic case though he is not going to make himself unpopular with anyone by saying whether he thinks selling arms to South Africa is the right way of doing it".[1] It has been suggested more openly by others, including General Hans Kruls, former chairman of the Netherlands Joint Chiefs of Staff and then editor of NATO's publication *NATO's Fifteen Nations,* that South Africa should become an "outside member" of that organization.

What are the prospects for success of Washington's policies?

If Suez was a turning point, a desperate effort by British and French imperialism and Israeli Zionism to turn back the Arab national-liberation movement by naked force, then the June war may well mark a similar decline for US imperialism—even if taking longer to unfold. In the late 50's and early 60's, US imperialism, in a world balance changing in favor of Socialism and national liberation, went over largely to the political and economic spheres and expanded its influence by trading on its non-colonial image and indirect ties, which gave it an initial advantage over its weaker partner/rivals. Subsequently, Washington's resort to political-military force to overcome the adverse trend, although understandable in the mid-60's, is less credible at the beginning of the 70's, and would appear to be a misreading of the present-day all-round balance of forces. Moreover, the United States-Israel relationship and the United States-South Africa and -Portugal ties may not prove a sufficient screen to ward off a polarization of forces against American imperialism.

Indications are that Washington, nevertheless, is continuing a policy of political-military force, especially through junior partners against Arabs in the north and Black Africans in the south, with the fate of the tropical African countries closely bound up with the outcome of the anti-imperialist struggle in both poles of the continent. In

---

[1] *The Economist,* January 9, 1971. This conservative organ suggests that a "political price" be paid by South Africa: "an easing of the banning system, more money for African welfare, the release of a few prisoners."

the long term, however, the African and world forces will more and more press US imperalism—despite reluctance—to downgrade its objectives or, at the very least, to lower the intensity of struggle by going over to greater political-economic or socio-economic emphasis as was done over a decade ago. But, if reaction in characteristic fashion delays too long in making this transition, then it may risk an even earlier loss of various sphere-of-influence structures—colonialism and neocolonialism.

V. AFTERWORD

Rather than update the text of the English edition, or append a chronicle of US African policy since 1971-1972, which becomes an endless task in analysis since events habitually overtake the publication process, it would seem of greater value to point out and illustrate in this added chapter key continuing tendencies and their implications.

Of central influence upon US foreign policy and the use or threat of military force has been the changed world balance, which is making imperialist reliance on advanced technology against popular movements much less applicable than in previous history. This was documented or borne out by at least three recent major events.

First, the "Agreement on Ending the War and Restoring the Peace in Vietnam" signed in Paris on January 27, 1973. After Vietnam, wrote James Reston of the *New York Times,* for example, it could no longer be taken for granted that big guys always lick little guys, that money and machines are decisive in war, and that small states would rather surrender than risk the military might of the United States.

Second, the "Agreement between the USSR and the USA for the Prevention of Nuclear War" signed in Washington by General Secretary Brezhnev and President Nixon on June 22, 1973. Significantly, the renunciation of the use or threat of force explicitly applied not only to the two signatory parties and their allies, but extended to other countries as well. This initial step also pointed the way to the institutionalization of political settlement of disputes between countries as a norm of international relations.

Third, the eruption of hostilities between the Arab countries and Israel on October 6, 1973, which exemplified the danger of not regulating a simmering and explosive conflict. Nevertheless this opened the way for political regulation. In sub-Sahara, too, there were continued political-military victories of national liberation movements in the Portuguese colonies, and even the political assassination of that outstanding leader Amilcar Cabral in January 1973 failed to prevent the establishment of an independent Guine-Bissau in September of the same year and its recognition by the UN.

The untenability of the imperialist "no war, no peace" policy and inevitability of a changed regional balance in the Middle East had been recognized in most quarters as being essentially a question of time. In this respect, however, western and Israeli government officials and specialists almost unanimously had deprecated the imminence or effectiveness of a new Arab-Israeli round, which finally took place. "Finally"—but really after only a short period of time even at the accelerated pace of contemporary history! As recently as July 28, 1973, the US delegate in vetoing a Security Council Resolution deploring the Israeli failure to pull out of occupied Arab lands had "blocked the way" according to Egypt's President, "for the attainment of a just political settlement." Indeed, until the very outbreak of October hostilities, the core of US imperialist strategy was major reliance on Tel Aviv's military machine, and secondarily political maneuver, in the expectation that the Arabs would be forced eventually to capitulate.

But Tel Aviv and Washington had underestimated a whole gamut of strength factors which altered the balance and made possible an Arab political-military and moral victory. Even in the strictly (if that is conceivable) military sphere, an Israeli "invincibility" doctrine based on air and armor superiority to hold occupied territory led, in fact, to an irredentist war with such high attrition rates to the aggressor's technology from anti-aircraft and anti-tank weapons as to cause many western military specialists to conclude that the infantry again had become the "queen of battle." For helping the Arab armies to achieve the necessary capability to withstand Tel Aviv's expansionist policy, President Sadat, Hafez Assad and other Arab leaders thanked the Soviet Union in particular.

Politically, the October war showed the force of Arab determination to regain seized lands and secure the legitimate national rights of the Palestinian Arabs. It proved the effectiveness of the solidarity of liberation movements with the Socialist community. Political support from Africa (most of which had severed ties with Israel before or soon after October 6), India and other non-aligned countries also helped to provide moral and diplomatic support to the Arabs and to isolate aggressive imperialism and Zionism on the international plane. The usefulness of growing world détente in preventing a widening of hostilities and in providing the medium for regulating a regional war was convincingly demonstrated. This, despite the US military alert of October 25-26, which rather than intimidating its adversaries, had the effect of exacerbating differences between the United States and western Europe and between imperialism and the third world.

There have been explanations, analyses and apologies for the war and its outcome. Thus, for example, Tel Aviv and Washington have attributed their surprise at the unexpectedness of the October war to subjective error in evaluating information—in effect, an inability or unwillingness to recognize an objectively changing balance since June 1967. Understandably, the character of reaction to live in the past tends to project a continuation of a previous balance of forces to the present. The "subjective" error, fundamentally, lies in not fully grasping the fact that the day has passed when imperialism can repeat nineteenth century foreign conquest. To be sure, even though the superior military force of reaction can impose the will of one class or country upon another, this by no means solves the underlying composite of socio-economic and national questions, which urgently press for radical solution. Therefore, when not in historical context, such battles as are won today can be only of a very limited or temporary nature.

One might have expected the political-military factor, in the Arab cause, which brought about the October 22 Security Council Resolution and ceasefire, to be of sufficient weight for Washington to bring its influence to bear on Tel Aviv to disgorge itself of Arab lands. An important role was also played by the additional Arab pressure on Washington which involved the use of oil as a political weapon and the ensuing oil crisis, the embargo against the United States and Netherlands, and, the general unwillingness of western Europe and Japan to subordinate national interests to the US-dominated multi-national companies. These factors *did* have the effect of bringing about negotiations at Geneva, and Israeli-Egyptian disengagement as a first step toward a political settlement.

At the same time, however, the newly improved bargaining position of the Afro-Arab states which had enabled them to step up a decade-long struggle for higher prices on and greater control of their long-exploited oil, as well as its employment as a political weapon to force adherence to international agreements, was not met passively by imperialism. Indeed, the experienced international cartels, which had produced some of the world's biggest billionaire fortunes on the basis of artificially depressed crude oil prices, could be expected to pass on the higher prices to consumers. As a result, the American oil monopolies are emerging with even higher prices than before, and simultaneously are seeking to avoid the onus of "price gouging" by inflaming domestic public opinion against the third world. But, beyond that, the threat by US officials to apply military force in the face of the Arab oil embargo was met by the producing countries with the counter-threat of mining and blowing up their oil installations. Similarly, the threat of raising US food export prices was generally received with scepticism because its consequent polarization of most of the world against imperialism would have redounded mainly to the latter's disadvantage.

In place of naked military force and crude economic blackmail, US imperialism is revealing a new range of political and diplomatic techniques for covering up its conflict of interests with the oil-producing states. In addition, economic projects have been proposed for mutually intertwining investment capital; the trading of US technology and arms for Arab politics and ideology; the promotion of private investment and erosion of the public sector, with its consequent socio-economic implications; and broad policies leading to inflation and currency devaluation, which erode not only the standard of living of the domestic working class and other unprivileged, but also the gains of the developing countries as well. It is questionable how effective each of these new individual forms of US neocolonialism may turn out.

In general, events themselves are pointing up the basic struggle of Africa—from north to south—against imperialism, and the logical consequence—the historic potential of growing African political consciousness and unity. Thus, the Tenth Assembly of the heads of state and government members of the Organization of African Unity (24-29 May, 1973) condemned colonialism, racism and Zionism, and the support given by the United States and other NATO countries to the reactionary regimes of South Africa and Rhodesia. The Assembly, furthermore, spoke out emphatically for cooperation with the Socialist community. Similarly, the Conference of non-aligned states in Algeria in September 1973 called on its participants to boycott Israel and to support the liberation forces in southern

Africa. At the same time, the Conference pointed to the necessity for world-wide solidarity in the face of "economic aggression," and for the establishment of full national control over natural resources and the right to nationalization; with the United States mentioned in this connection as the main imperialist opponent of the African states. In December 1973, the Arab summit conference in Algeria decided to cut off all Arab oil to South Africa, Rhodesia and Portugal—a further display of continental solidarity.

In sum, the third world is increasingly linking up the political, economic and military struggle against imperialism on a world scale. The non-aligned countries meeting in Algiers in March 1974 condemned US and Saigon violations of the Paris agreements, imperialist political and economic aid to the colonial and racist regimes, and economic blockade in Latin America. They urged a radical transformation of the imperialist structure of economic relations which is based on inequality, domination and exploitation. For even though economic progress in Africa and other developing continents is taking place, it is generally regarded as being too slow in view of world potentialities and in the face of imperialist obstacles.

This growing understanding of the forces and mechanisms at work, and awareness of the need for anti-imperialist unity on all levels, is undoubtedly the most effective guarantee of further success in Africa's struggle for progress.

March 1974

INDEX